Springer Labor Manual

Martin Holtzhauer

Basic Methods for the Biochemical Lab

First English Edition

23 Figures and 86 Tables

Springer

Dr. Martin Holtzhauer
Human GmbH Branch IMTEC
Robert-Rössle-Strasse 10
13125 Berlin
Germany
e-mail: m.holtzhauer@imtec-berlin.de

ISBN 3-540-19267-0 1st German edition Springer-Verlag Berlin Heidelberg New York 1988
ISBN 3-540-58584-2 2nd German revised edition Springer-Verlag Berlin Heidelberg New York 1995
ISBN 3-540-62435-X 3rd German revised edition Springer-Verlag Berlin Heidelberg New York 1997

Library of Congress Control Number: 2006922621

ISBN-10 3-540-32785-1 **Springer Berlin Heidelberg New York**
ISBN-13 978-3-540-32785-1 **Springer Berlin Heidelberg New York**

Springer is a part of Springer Science+Business Media
springer.com

Printed in Germany

Cover design: *design&production*, Heidelberg, Germany
Typesetting and production: LE-TEX Jelonek, Schmidt & Vöckler GbR, Leipzig, Germany

2/3141 YL 5 4 3 2 1 0 - Printed on acid-free paper

For Dorothea,
Susanne, and Christian

Preface

More than 20 years ago I started a collection of adapted protocols modified for special applications and checked for daily usage in the biochemical (protein) lab. Small "methods" within large papers or parts of chapters in special books, overloaded with theoretical explanations, were the basis. My imagination was a cookbook: Each protocol contains a list of ingredients and a short instruction (sometimes I was not very consequent, I beg your pardon!). I proposed this idea to some publishing houses, and in 1988 Springer-Verlag published the first edition of *Biochemische Labormethoden*. Interest and suggestions of numerous colleagues led to a second and third German edition, and now there seems to be an interest outside Germany, too. The contents and form of this cookbook are perhaps helpful for students, technicians, and scientists in biochemistry, molecular biology, biotechnology, and clinical laboratory.

Starting from the first edition, the aim of this book has been to provide support on the bench and a stimulation of user's methodological knowledge, resulting in a possible qualification of his/her experimental repertoire and, as a special request for the reader of this book, an improvement of the "basic protocols."

During my professional life I have received innumerable hints and special tips from a multitude of colleagues and co-workers. Their knowledge is now part of the present protocols and I give my thanks to them.

I especially acknowledge Mrs. Susanne Dowe, because without her support and helpful criticism, I never would have tried to make a further edition of these protocols.

Berlin, January 2006 *Martin Holtzhauer*

Table of Contents

Abbreviations . . . XVII

1 **Quantitative Methods** . . . 1
1.1 Quantitative Determinations of Proteins . . . 1
1.1.1 Lowry Protein Quantification . . . 2
1.1.1.1 Standard Procedure . . . 2
1.1.1.2 Modification by Sargent . . . 3
1.1.1.3 Micromethod on Microtest Plates . . . 4
1.1.1.4 Protein Determination in the Presence of Interfering Substances . . . 5
1.1.2 Bradford Protein Determination . . . 6
1.1.3 Protein Determination in SDS-PAGE Sample Solutions . . . 7
1.1.4 Protein Determination Using Amido Black . . . 8
1.1.5 BCA Protein Determination . . . 9
1.1.5.1 BCA Standard Procedure . . . 9
1.1.5.2 BCA Micromethod . . . 9
1.1.6 Kjeldahl Protein Determination . . . 10
1.1.7 UV Photometric Assay of Protein Concentration . . . 11
1.2 Quantitative Determination of Nucleic Acids . . . 13
1.2.1 Schmidt and Thannhauser DNA, RNA, and Protein Separation Procedure . . . 13
1.2.2 Orcin RNA (Ribose) Determination . . . 14
1.2.3 Diphenylamine DNA (Deoxyribose) Determination . . . 14
1.2.4 Quantitative DNA Determination with Fluorescent Dyes . . . 15
1.2.5 Determination of Nucleic Acids by UV Absorption . . . 16
1.3 Quantitative Phosphate Determinations . . . 17
1.3.1 Determination of Inorganic Phosphate in Biologic Samples . . . 17
1.3.2 Determination of Total Phosphate . . . 18
1.3.3 Phospholipid Determination . . . 18
1.4 Monosaccharide Determination . . . 19
1.5 Calculations in Quantitative Analysis . . . 20

2 **Electrophoresis** . . . 23
2.1 Polyacrylamide Gel Electrophoresis Systems . . . 23
2.1.1 Laemmli SDS-Polyacrylamide Gel Electrophoresis . . . 26
2.1.2 SDS-Polyacrylamide Gel Electrophoresis at Neutral pH (NuPAGE) . . . 31

2.1.3 SDS-Polyacrylamide Gel Electrophoresis According to Weber, Pringle, and Osborn 32
2.1.4 Urea-SDS-Polyacrylamide Gel Electrophoresis for the Separation of Low Molecular Weight Proteins 34
2.1.5 TRICINE-SDS-Polyacrylamide Gel Electrophoresis for Proteins and Oligopeptides in the Range of 1000–50 000 Daltons 35
2.1.6 SDS-Polyacrylamide Gel Electrophoresis at pH 2.4 36
2.1.7 Urea-Polyacrylamide Gel Electrophoresis for Basic Proteins at pH 2 37
2.1.8 Anodic Discontinuous Polyacrylamide Gel Electrophoresis (Native PAGE) 38
2.1.9 Cathodic Discontinuous Polyacrylamide Gel Electrophoresis (Native PAGE) 39
2.1.10 Affinity Electrophoresis 40
2.1.11 Two-Dimensional Polyacrylamide Gel Electrophoresis (2D-PAGE; IEF followed by SDS-PAGE) 41
2.1.11.1 First Dimension: Isoelectric Focusing (IEF) 42
2.1.11.2 Second Dimension: SDS-PAGE (Acrylamide Gradient Gel) 44
2.2 Agarose and Paper Electrophoresis 45
2.2.1 Non-denaturating Nucleic Acid Electrophoresis 45
2.2.2 Denaturating Nucleic Acid Electrophoresis 46
2.2.3 Identification of Phosphoamino Acids (Paper Electrophoresis) . 48
2.3 Aid in Electrophoresis 49
2.3.1 Marker Dyes for Monitoring Electrophoresis 49
2.3.1.1 Anodic Systems 49
2.3.1.2 Cathodic Systems 49
2.3.2 Marker Proteins for the Polyacrylamide Gel Electrophoresis ... 50
2.3.3 Covalently Colored Marker Proteins 52
2.4 Staining Protocols 53
2.4.1 Staining with Organic Dyes 53
2.4.1.1 Amido Black 10 B 54
2.4.1.2 Coomassie Brilliant Blue R250 and G250 54
2.4.1.3 Coomassie Brilliant Blue R250 Combined with Bismarck Brown R 55
2.4.1.4 Fast Green FCF 55
2.4.1.5 Stains All 56
2.4.1.6 Staining of Proteolipids, Lipids, and Lipoproteins.... 56
2.4.2 Silver Staining of Proteins in Gels 56
2.4.2.1 Citrate/Formaldehyde Development 57
2.4.2.2 Alkaline Development 58
2.4.2.3 Silver Staining Using Tungstosilicic Acid 58
2.4.2.4 Silver Staining of Proteins: Formaldehyde Fixation .. 59
2.4.2.5 Silver Staining of Glycoproteins and Polysaccharides . 60
2.4.2.6 Enhancement of Silver Staining 60
2.4.2.7 Reducing of Silver-Stained Gels 61
2.4.3 Copper Staining of SDS-PAGE Gels 61

2.4.4 Staining of Glycoproteins and Polysaccharides in Gels 62
2.4.4.1 Staining with SCHIFF's Reagent (PAS Staining) 62
2.4.4.2 Staining with Thymol 63
2.4.5 Staining of Blotted Proteins on Membranes 63
2.4.5.1 Staining on Nitrocellulose with Dyes 63
2.4.5.2 Staining on Nitrocellulose with Colloidal Gold 64
2.4.5.3 Staining on PVDF Blotting Membranes with Dyes ... 65
2.5 Electroelution from Gels 66
2.5.1 Preparative Electroelution of Proteins from Polyacrylamide Gels 66
2.5.2 Removal of SDS 67
2.5.3 Electrotransfer of Proteins onto Membranes (Electroblotting; Western Blot): Semi-dry Blotting 68
2.5.4 Immunochemical Detection of Antigens After Electrotransfer (Immunoblotting) 70
2.5.4.1 Detection Using Horseradish Peroxidase (HRP) 72
2.5.4.2 Detection Using Alkaline Phosphatase (AP) 73
2.5.5 Chemiluminescence Detection on Blotting Membranes 74
2.5.5.1 Chemiluminescence Using HRP 74
2.5.5.2 Chemiluminescence Using AP 74
2.5.6 Carbohydrate-Specific Glycoprotein Detection After Electrotransfer 75
2.5.7 General Carbohydrate Detection on Western Blots 76
2.5.8 Affinity Blotting 77
2.5.9 Transfer of Nucleic Acids (SOUTHERN and Northern Blot) 78
2.6 Drying of Electrophoresis Gels 79
2.7 Autoradiography of Radioactive Labeled Compounds in Gels 80

3 Chromatography 83
3.1 Thin-Layer Chromatography 83
3.1.1 Identification of the N-terminal Amino Acid in Polypeptides (TLC of Modified Amino Acids 83
3.1.2 Thin-Layer Chromatography of Nucleoside Phosphates 85
3.1.3 Gradient Thin-Layer Chromatography of Nucleotides 85
3.1.4 Identification of Phosphates on TLC Plates 87
3.1.5 Lipid Extraction and TLC of Lipids 88
3.2 Hints for Column Chromatography of Proteins 89
3.3 Gel Permeation Chromatography (GPC; Gel Filtration, GF; Size-Exclusion Chromatography, SEC) 93
3.3.1 Selection of Supports 96
3.3.2 Filling of a Gel Filtration Column 97
3.3.3 Sample Application and Chromatographic Separation (Elution) 97
3.3.4 Cleaning and Storage 98
3.3.5 Determination of Void Volume V_0 and Total Volume V_t 99
3.3.6 Removing of Unbound Biotin After Conjugation by Gel Filtration ("Desalting") 99
3.4 Ion Exchange Chromatography (IEC) 102
3.4.1 Preparation of Ion Exchange Supports 103

3.4.2 Capacity Test ... 104
3.4.3 Sample Application ... 104
3.4.4 Elution ... 105
3.4.5 Cleaning and Regeneration ... 105
3.4.6 High-Performance Ion Exchange Chromatography (HPIEC) of Mono- and Oligosaccharides ... 106
3.5 Hydrophobic Interaction Chromatography (HIC) ... 107
3.5.1 Capacity Test ... 107
3.5.2 Elution ... 108
3.5.3 Regeneration ... 108
3.5.4 Analytical HPLC of Hapten-Protein Conjugates ... 108
3.6 Affinity Chromatography (AC) ... 109
3.6.1 Cyanogen Bromide Activation of Polysaccharide-Based Supports ... 113
3.6.1.1 Determination of the Degree of Activation ... 114
3.6.2 Coupling to Cyanogen Bromide-Activated Gels ... 114
3.6.2.1 Quantitative Determination of Coupled Diamine Spacers with 2,4,6-Trinitrobenzene Sulfonic Acid ... 115
3.6.2.2 Quantitative Determination of Immobilized Protein ... 116
3.6.2.3 Immobilization of Wheat Germ Agglutinin ... 116
3.6.2.4 Affinity Purification of HRP ... 117
3.6.2.5 Affinity Chromatography of Immunoglobulins on Immobilized Antibodies (Immunoaffinity Chromatography, IAC) ... 117
3.6.2.6 Affinity Chromatography of Rabbit IgG on Protein-A Supports ... 118
3.6.3 Activation of Sepharose with Epichlorohydrin ... 119
3.6.3.1 Determination of Epoxy Residues ... 119
3.6.4 Immobilization of Monosaccharides (Fucose) ... 119
3.6.5 Activation with Divinylsulfone ... 120
3.6.6 Coupling of Reactive Dyes to Polysaccharides (Dye-Ligand Chromatography) ... 121
3.6.7 Covalent Coupling of Biotin (Biotin-Avidin/Streptavidin System) ... 121
3.6.8 Metal Chelate Chromatography of Proteins Containing His_6-Tag ... 123
3.7 Concentration of Diluted Protein Solutions ... 124
3.7.1 Acidic Precipitation ... 124
3.7.2 Salting Out ... 124
3.7.3 Precipitation Using Organic Substances ... 125
3.7.4 Lyophilization (Freeze Drying) ... 126
3.7.5 Ultrafiltration ... 127

4 Immunochemical Protocols ... 129
4.1 Conjugation of Haptens (Peptides) to Carrier Proteins ... 129
4.1.1 Activation of Proteins with Traut's Reagent Yielding Proteins with Additional Free SH Groups ... 132
4.1.2 Conjugation of MCA-Gly Peptides to SH-Carrying Proteins ... 132

4.1.3 Conjugation of Sulfhydryl Peptides Using 4-(N-Maleimidomethyl)-Cyclohexane-1-Carbonic Acid N-Hydroxysuccinimide Ester (SMCC) 133
4.1.4 β-Galactosidase-Immunoglobulin Conjugate (Coupling via SH Groups) 134
4.1.4.1 Enzyme Reaction of β-Galactosidase 134
4.1.5 Carbodiimide Coupling of Peptides to Carrier Proteins with 1-Ethyl-3-(3-Dimethylaminopropyl)-Carbodiimide (EDAC, EDC) 134
4.1.6 Conjugation of Horseradish Peroxidase (Glycoproteins) by Periodate Oxidation 135
4.1.7 Conjugation of Peptides to Carrier Proteins Using Glutaraldehyde (Two-Step Procedure) 136
4.1.8 Conjugation of HRP to Antibodies with Glutaraldehyde 137
4.1.9 Alkaline Phosphatase-Immunoglobulin Conjugate (Glutaraldehyde Protocol) 138
4.1.9.1 Enzymatic Reaction of Alkaline Phosphatase from Calf Intestine 138
4.1.10 Labeling of Immunoglobulins with Fluorescent Dyes 138
4.1.11 Protein-Colloidal Gold Conjugates 141
4.1.11.1 Preparation of Colloidal Gold Sol 141
4.1.11.2 Adsorption of Protein to Colloidal Gold 142
4.2 Immunization of Laboratory Animals 143
4.3 Ammonium Sulfate Fractionation of Immunoglobulins 144
4.4 Removal of Unspecific Immunoreactivities 146
4.4.1 Preparation of Tissue Powder (Liver Powder) 148
4.5 Preparation of Egg Yolk IgY Fraction 148
4.6 Antibody Fragmentation 149
4.6.1 F(ab′)2 Fragments from IgG 149
4.6.2 Fab′ Fragments (Rabbit) 150
4.6.3 Fab Fragments (Rabbit) 150
4.7 HEIDELBERGER Curve (Precipitin Curve) 150
4.8 OUCHTERLONY Double-Radial Immunodiffusion 151
4.8.1 Purification of Agar 151
4.8.2 Preparation of Slides 151
4.8.3 Immunodiffusion 152
4.8.4 Visualization of the Precipitin Lines 152
4.9 Immunoprecipitation of Antigens 153
4.10 Immunoelectrophoresis 154
4.11 Counterelectrophoresis 155
4.12 Dot-Blot Assay 156
4.13 Enzyme Immunosorbent Assay (EIA, ELISA) 157
4.13.1 Indirect EIA with HRP Conjugate 158
4.13.2 Determination of Enzyme Activity by ELISA 159
4.13.3 Isotype Determination by EIA (AP Conjugate) 160

5 **Centrifugation** 161
5.1 Speed vs Centrifugal Force Graphs 161
5.2 Differential Centrifugation 164
5.3 Density Gradient Centrifugation 165
5.3.1 Pre-formed Discontinuous Gradient Centrifugation: Isolation of Liver Cell Nuclei 166
5.3.2 Sucrose Gradient Centrifugation: Preparation of Surface Membranes (Sarcolemma, SL) of Heart Muscle Cells 167
5.3.2.1 Determination of a Marker Enzyme: Ouabain-Sensitive Na,K-ATPase 172
5.3.2.2 Receptor Determination: DHP Binding Sites on Surface Membranes 173
5.3.2.3 Determination of the Dissociation and Association Kinetics of the DHP Receptors 174
5.3.3 RNA Separation by Non-Denaturating Sucrose Density Gradient Centrifugation 175
5.3.4 Denaturating RNA Gradient Centrifugation 176
5.3.5 Isopycnic Centrifugation 177
5.3.5.1 Purification of High Molecular Weight DNA in CsCl Gradients 177
5.3.5.2 Cell Fractionation Using Percoll 178
5.3.5.3 Preparation of Human Lymphocytes 179

6 **Radioactive Labeling** 181
6.1 Radioactive Decay 182
6.2 Decay Tables for 32-Phosphorus, 35-Sulfur, and 125-Iodine 183
6.3 Enzymatic [^{32}P]-Phosphate Incorporation into Proteins 185
6.4 Iodination with [^{125}I]-Iodine Reagents 187
6.4.1 Chloramine-T Protocol 187
6.4.2 Iodination with Bolton–Hunter Reagent 188
6.5 Scintillation Cocktails for Liquid Scintillation Counting 188

7 **Buffers** 191
7.1 Theoretical Considerations 191
7.2 Plot for Buffer Calculations 198
7.3 pH Indicators 199
7.4 Buffer Recipes 199
7.4.1 Commonly Used Buffers 201
7.4.2 Buffers and Media for Tissue and Cell Culture and Organ Perfusion 204
7.4.3 pH Calibration Buffers 206
7.4.4 Volatile Buffers 207

8 **Tables** 209
8.1 Concentration Units 209
8.2 Conversion Factors for SI Units 210
8.3 Data of Frequently Used Substances 212
8.4 Protein Data 216

8.5 Protease Inhibitors 221
8.6 Single-Letter Codes and Molecular Masses of Amino Acids 222
8.7 Spectroscopic Data of Nucleotides 225
8.8 Detergents ("Surfactants") 225
8.9 Refractive Index and Density of Sucrose Solutions 228
8.10 Ammonium Sulfate Saturation Table 229
8.11 Diluted Solutions 231
8.12 Mixture Rule 232

9 Statistics and Data Analysis 233
9.1 Statistical Equations 233
9.1.1 Mean and Related Functions 233
9.1.2 Correlation: Linear Regression 234
9.1.3 The *t*-test (Student's Test) 236
9.2 Data Analysis 237
9.2.1 Receptor–Ligand Binding 237
9.2.2 Enzyme Kinetics 240
9.2.3 Determination of Molecular Mass by SDS-PAGE 243
9.3 Diagnostic Sensitivity and Specificity 244
9.4 Software for the Lab 244
9.4.1 Data Analysis and Presentation 245
9.4.2 Software for Statistics 245
9.4.3 Other Software 245
9.4.4 Selected Internet Links 246

Subject Index 247

Abbreviations

A_{280}	absorption of light with wavelength 280 nm
$A_{280}^{1\%}$	absorption coefficient of a 1% solution at 280 nm
Ag	antigen
Ab	antibody
AP	alkaline phosphatase
bp	base pairs (of nucleic acids)
BSA	bovine serum albumin
%C	percent cross-linker of total amount T of acrylamide monomers
cAMP	cyclic AMP
cc	constant current
cv	constant voltage
D	Dalton (relative molecular mass)
ddH_2O	ultrapure (double distilled/reverse osmosis) water
DMF	dimethylforamide
DMSO	dimethylsulfoxide
dpm	decays per minute
DTE	erythro-1,4-dimercapto-2,3-butanediol (dithioerythreitol, CLELAND's reagent)
DTT	threo-1,4-dimercapto-2,3-butanediol (dithiothreitol, CLELAND's reagent)
ε_{280}	molar absorption coefficient at 280 nm
EDTA	ethylenediamintetraacetic acid, disodium salt
EGTA	ethyleneglycol-bis(N,N,N′,N′-aminoethyl) tetraacetic acid
EIA	enzyme-linked immunoassay (ELISA, enzyme-linked immunosorbent assay)
g	relative centrifugal force ($1\ g = 9.81\ m \cdot s^{-2}$)
g_{av}	g at mean distance from the rotor center
g_{max}	g at maximal distance from the rotor center
HPLC	high-performance liquid chromatography
HRP	horseradish peroxidase
I	ionic strength
Ig	immunoglobulin (e.g., IgG – immunoglobulin G)
kD	kiloDalton (10^3 D)
KLH	keyhole limpet hemocyanin
M	molar (moles per liter)
M_r	relative molar mass
mAb	monoclonal antibody
mol-%	molecules per 100 molecules/moles per 100 moles
N	normal (vales per liter)
NEM	N-ethylmaleinimide

O.D.	optical density
PAGE	polyacrylamide gel electrophoresis
PBS	phosphate-buffered saline
pI	isoelectric point
pK	negative common logarithm of equilibrium constant
PMSF	phenylmethanesulfonyl fluoride
PVDF	polyvinylidene difluoride
R_f	relative migration distance
rpm	revolutions per minute
RT	room temperature
ϱ	density, specific gravity
SD	standard deviation of mean
SDS	sodium dodecylsulfate
Soln.	solution
%T	percentage (w/w) of whole acrylamide (acrylamide + cross-linker) in a PAGE gel
TBS	Tris-buffered saline
TCA	trichloroacetic acid
Tris	tris(hydroxymethyl) aminomethane
UV	ultraviolet light
v/v	volume for (total) volume
w/v	weight for (total) volume
w/w	weight for (total) weight

1 Quantitative Methods

1.1 Quantitative Determinations of Proteins

The quantitative estimation of proteins is one of the basic requirements in biochemistry. In reviewing the biochemical literature for methods of fast and sensitive determination of the amount of protein, the large variety of proteins becomes evident, since the amount of protocols for quantitative protein seems to be innumerable.

Proteins, from many points of view, are much more complex than, for example, nucleic acids. As a result, it has been difficult to give laboratory protocols that can be applied to proteins in general; however, in most cases the specialized protocols may be reduced to a few basic methods. But if a protein becomes pure or some of its unique properties are of special interest, another analytical method has to be used. Nevertheless, accurate quantitation of the amount of protein during the steps of protein preparation is the only valid way to evaluate the overall value of a procedure.

The following protocols are based on distinct properties of proteins; therefore, exact information is only possible if a heterogeneous protein mixture is compared with a universal standard protein. The best way would be to take a defined sample of the protein to be analyzed. So the difficulties start with the selection of the standards, because it is well known how difficult it is to prepare a protein that fulfills the criteria of analytical chemistry.

It is very often observed that during a purification process the differences increase between the real amounts of a protein and the values obtained by any method, e.g., total enzyme activity, because the measured signal produced by a protein mixture differs from that of a pure protein. Furthermore, the amount of a given protein determined by a distinct protocol differs from the expected amount by portioning, as shown in Table 1.1. To avoid additional mistakes with the already uncertain process, the protein estimation method should not be changed during a purification process.

With these difficulties kept in mind, any protein may be estimated by one of the given protocols. Absolute statements, such as "... the prepared, pure product has a specific activity of ... units per milligram of protein ..." should be made with caution.

References

Stoscheck CM (1990) Meth Enzymol 182:50
Sapan CV, Lundblad RL, Price NC (1999) Biotechnol Appl Biochem 29:99

Table 1.1. Comparison of various quantitation methods[a]

Protein	Lowry	Bradford	BCA	Fluorescence
α-Casein				0.91
BSA	1.00	1.00	1.00	1.00
Calf thymus histones	1.24	1.10		1.15
Chymotryp-siongen A	1.52	0.58	0.99	
Cytochrome c	1.39	1.20	1.11	
γ-Globulin	1.07	0.46	0.95	
IgG (human)			1.03	
IgG (mouse)			1.23	0.79
Lysozyme	1.54	1.00		0.97
Myoglobin	0.84	1.38	0.92	0.91
Ovalbumin	0.93	0.52	1.08	0.97
Ribonuclease A	1.28	0.68		
Soybean trypsin inhibitor	0.83	0.66		
Trypsin	1.34	0.34		
Mean ± SD	*1.18 ± 0.26*	*0.81 ± 0.34*	*1.04 ± 0.10*	*0.96 ± 0.11*

[a] Estimated for highly purified proteins; relative to BSA. Data from: Peterson GL (1983) Meth Enzymol 91:95; Pierce (1996) Protein assay technical protocol; and Invitrogen/Molecular Probe Quant-iT Technical Bulletin (2004)

1.1.1 Lowry Protein Quantification

1.1.1.1 Standard Procedure

This protocol is slightly modified, with respect to the original paper by Lowry et al., to work with smaller volumes. The Folin phenol method (Lowry protocol; Table 1.2) is useful in the widest variety of experimental applications and is also the least variable with different proteins. It is noted that this method, which uses the oxidation of aromatic amino acids, is easily disturbed by a lot of substances, which are components of the buffer. As a control an aliquot of the protein-free buffer in the same volume as the protein-containing sample has to be taken as blank[1].

Since the reaction conditions may differ from experiment to experiment and the standard curve is not linear, a couple of standards with different amounts of protein between 0 and 100 µg should be measured in each analysis. For most purposes a stock solution of

[1] A detailed discussion of Folin–Ciocalteu's phenol protein determination method, especially with respect to possible disturbances and troubles and in comparison with the Bradford method, is given by Peterson (1996) loc. cit.

ovalbumin (Ova) or bovine serum albumin (BSA) in 0.1% SDS (w/v) is suitable. This solution can be stored in the refrigerator for several weeks.

Solutions/Reagents

A 20 g Na_2CO_3 (anhydrous) in 1000 ml 0.1 N NaOH
B 1.0 g $CuSO_4 \cdot 5H_2O$ in 100 ml ddH_2O
C 2.0 g potassium-sodium tartrate in 100 ml ddH_2O
D mix 1 vol. B and 1 vol. C, and then add 50 vol. A
E FOLIN–CIOCALTEU's phenol reagent (stock), 1 + 1 diluted with ddH_2O
Standard 5.0 mg/ml ovalbumin or BSA, 0.1% SDS (w/v) in ddH_2O

Standard protocol

Table 1.2. LOWRY standard protocol

	Blank	Standard (ml)	Sample
	–	Max. 0.1	–
	–	–	Max. 0.1
H_2O	0.1	to 0.1	to 0.1
Soln. D	2.0	2.0	2.0
Mix, incubate for 5–10 min at RT			
Soln. E	0.2	0.2	0.2
Mix, incubate for 30–45 min at RT, read at 700 nm			

Make samples, blank, and standards at least in duplicates, and measure in a spectrophotometer at 700–750 nm.

Half-micro protocol

Especially for small amounts of protein, reduce the volumes: 0.1 ml of 0.1% SDS in ddH_2O are added to 0.10 ml of sample, and then add 1.0 ml Soln. E, and 5 min thereafter add 0.1 ml Soln. D. Measure after 30–45 min.

Prepare the standard curve in the range between 0 and 30 µg of protein. Since the standard curve in this range is nearly linear, it is possible to take a factor F, which can be estimated at that time when solution D is used for the first time.

$$\mu g \text{ Protein}/100\ \mu l = \left(A_{Sample} - A_{Blank}\right) \cdot F$$

Mix suspensions of membrane proteins, cell homogenates, etc., with an equal volume of 0.1 NaOH to get a homogenous solution.

For estimation of proteins covalently bound to chromatographic matrices hydrolyze the sample for 6 h at 37 °C in solution D. After centrifugation, use an aliquot for protein determination.

1.1.1.2 Modification by SARGENT

A 50-fold increase in sensitivity with respect to the LOWRY standard protocol was described by SARGENT. It is possible to estimate 0.1–1 µg protein and 4–40 µg/ml, respectively.

Solutions/Reagents

A $20\ mM\ CuSO_4$, 40 mM citric acid, 0.1 mM EDTA
B $0.4\ M\ Na_2CO_3$, 0.32 M NaOH
C mix 1 vol. freshly prepared A with 25 vol. freshly prepared B
D FOLIN–CIACALTEU's phenol reagent (stock), 1 + 1 diluted with ddH_2O
E 60 µg/ml malachite green in 0.1 M sodium maleate buffer, pH 6.0, 1 mM EDTA

Measure at 690 nm immediately after addition of solution E. The assay may be done in a microtest plate (Table 1.3).

Micro assay

Table 1.3. LOWRY microassay

	Blank	Standard (µl)	Sample
Buffer	Max. 15	–	–
Standard	–	Max. 15	–
Sample	–	–	Max. 15
H_2O	to 15	to 15	to 15
Solution C	15	15	15
	Mix, incubate for 15 min at RT		
Solution D	3	3	3
	Mix, incubate for 30–45 min at RT		
Solution E	180	180	180

Detergents, e.g., SDS, at elevated concentrations strongly disturb the test. If at high blank level the difference between blank and sample is too small, this interference should be omitted by an extraction of the detergent (cf. Protocol 1.1.2 and 1.1.4).

Prior to the addition of Soln. E, extract the sample twice with 1 ml ethyl ether each. Remove the ether by aspiration after centrifugation; remove remaining ether in the aqueous phase with a SpeedVac. Prepare the standard curve in the range between 0 and 1 µg BSA. This extraction of detergents is not allowed to be done in a microtest plate.

References

Lowry OH, Rosebrough NJ, Farr AL, Randall RL (1951) J Biol Chem 193:265
Sargent MG (1987) Anal Biochem 163:476

1.1.1.3 Micromethod on Microtest Plates

Between 0.5 and 80 µg of protein (equivalent to 20–1600 µg/ml) may be estimated in a microtest plate (96-well plate, flat bottom).

Solutions/Reagents

A 20 g Na_2CO_3 (anhydrous) in 1000 ml 0.1 N NaOH
B 1.0 g $CuSO_4 \cdot 5H_2O$ in 100 ml ddH_2O

C 2.0 g potassium-sodium tartrate (Seignette salt) in 100 ml ddH_2O

D mix 1 vol. B and 1 vol. C, and then add 50 vol. A

E FOLIN–CIOCALTEU's phenol reagent (stock), 1 + 1 diluted with ddH_2O (Tables 1.4, 1.5)

Standard 2.0 mg/ml BSA in 0.1 N NaOH (stable at 2–8 °C for several months)

Dilute the sample with sodium hydroxide to a final concentration of about 0.1 moles NaOH/l and to an amount of protein within the measuring range.

Table 1.4. Dilution protocol of the microassay (0–40 µg; FOLIN method)

Standard (µl)	0.1 N NaOH (µl)	Protein per assay (µg)	(µg/ml)
0	100	0	0
10	90	5	200
20	80	10	400
30	70	15	600
40	60	20	800
50	50	25	1000
60	40	30	1200
80	20	40	1600

Table 1.5. Protocol of micromethod (FOLIN method)

Sample and standard, respectively	25 µl
Soln. D	250 µl
	Mix on a shaker for 5–10 min at RT
Soln. E	25 µl
	Mix on a shaker for 30–45 min at RT
	Measure with an EIA reader at 620 nm

1.1.1.4 Protein Determination in the Presence of Interfering Substances

If a sample contains a larger amount of interfering substances, i.e., the blank gives a high value, these substances may be removed according to this protocol. But some detergents, such as digitonine, prevent the precipitation of the proteins.

Solutions/Reagents

A 0.15% sodium deoxycholate (w/v) in ddH_2O

B 72% trichloroacetic acid (w/v) in ddH_2O

C 1% $CuSO_4$ (w/v) in ddH_2O

D 2% sodium-potassium tartrate (w/v) in ddH_2O

E 3.4% sodium carbonate (anhydrous) (w/v) in 0.2 N NaOH

F 10% SDS (w/v) in ddH_2O
G Mix just before use Soln. C, D, E, and F in a ratio of 1:1:28:10
H FOLIN–CIOCALTEU's phenol reagent (stock), diluted 1 + 3 with ddH_2O

Five to 100 µg of protein and standard, respectively, are diluted with ddH_2O to 1.0 ml. After that, add 0.1 ml of Soln. A. After further 10 min at RT add 0.1 ml of Soln. B. Mix the solution well, centrifuge the samples with 3000 × g at RT 15 min later.

Resolve the precipitate in 1.0 ml ddH_2O, and then add 1.0 ml Soln. G. Now the precipitate should be resolved completely. Add 0.5 ml of Soln. H after further 10 min, mix well and read 30–45 min later at 700 nm.

In a microassay after centrifugation, the volumes can be reduced to 1/5.

References

Peterson GL (1983) Meth Enzymol 91:95

1.1.2 BRADFORD Protein Determination

BRADFORD assay protocol (Table 1.6) is less time-consuming and is little, or not at all, disturbed by most buffers and reducing substances. On the other hand, detergents such as, for example, deoxycholate or Triton X100 make trouble because they form coarse precipitates in the strong acidic reagent, and this method also gives false results if the sample is microheterogeneous, as observed in the case of some membrane protein preparations. The SDS interferes strongly at concentration above 0.2%[2]. The blank is mostly high, but there is no influence on the measurement.

Solutions/Reagents

A 0.1 g Coomassie Brilliant Blue G 250 (C.I. 42655[3]) are dissolved in 50 ml 50% ethanol (v/v). After that, 100 ml of 85% phosphoric acid are added and made up with ddH_2O to a total volume

Table 1.6. BRADFORD assay protocol

	Blank (ml)	Standard (ml)	Sample (ml)
	–	Max. 0.1	–
	–	–	Max. 0.1
H_2O	0.1	to 0.1	to 0.1
Solution B	2	2	2

[2] The influence of detergents was examined, for example, by FRIEDENAUER and BERLET (1989) Anal Biochem 178:263

[3] C.I. Color Index, international system for identification of chemical dyes

of 250 ml. This stock solution should be prepared about 4 weeks before use. It is stable for several months at 2–8 °C.

B Dilute 1 vol. Soln. A with 4 vol. ddH_2O and filter the mixture before use.

Standard: 5.0 mg/ml in 0.1% SDS (w/v)

The protein solution (standard and sample) should contain 10–100 µg protein.

After pipetting the solutions, mix them well and read the absorption at 590 nm 5 min later. The protein-dye complex is stable for a longer period.

Prepare the standard curve by a serial dilution of a BSA stock between 0 and 125 µg.

Half-micro assay on microtest plates

For samples with less than 50 µg of protein, the protocol is modified as follows: complete up to 50 µl sample with ddH_2O to a total volume of 800 µl. Add 200 µl Soln. A and mix thoroughly. Measure the absorption at 590 nm after 5 min.

Since the standard curve is nearly linear in the range up to 50 µg, a constant for Soln. A can be determined and used for the whole lot of Soln. A.

The use of disposable plastic cuvettes is recommended. If glass cuvettes are used, remove adhered protein-dye complex on the walls with 96% ethanol or methanol.

References

Bradford MM (1976) Anal Biochem 72:248

1.1.3 Protein Determination in SDS-PAGE Sample Solutions

Some components of sample buffers, e.g., Tris or 2-mercaptoethanol, disturb most of the (chemical) protein determinations. If the lanes of an electrophoresis should be compared quantitatively, if a UV measurement of the sample is impossible, and if the sample contains enough material, the protein content in electrophoresis sample buffer can be measured using the following protocol.

Solutions/Reagents

A ***electrophoresis sample buffer***: 62.5 mM Tris · HCl, pH 6.8, 2% SDS (w/v), 5% 2-mercaptoethanol (v/v), 10% sucrose (w/v)

B 0.1 M potassium phosphate buffer, pH 7.4 (cf. Table 1.7)

***Important!** The use of potassium phosphate is essential!*

C dissolve 50 mg Coomassie Brilliant Blue G 250 in 50 ml ddH_2O and add 50 ml 1 M perchloric acid

Standard: 5.0 mg/ml in 0.1% SDS (w/v)

Fill 20 µl of sample in buffer A up with ddH_2O to 50 µl. After that, 0.45 ml Soln. B are added. Vortex the solutions and centrifuge after 5–10 min with 1500–2000 × g at RT for 10 min.

Mix 0.25 ml of the clarified supernatant with 2.75 ml of Soln. C and read in a photometer at 620 nm. Prepare the standard curve in the range from 10 to 100 µg. The blank is made from 20 µl of Soln. A.

References

Zaman Z, Verwilghan RL (1979) Anal Biochem 100:64

1.1.4 Protein Determination Using Amido Black

Solutions/Reagents

A 0.1% Amido Black 10 B (C.I. 20470) (w/v) in 30% methanol (v/v), 70% acetic acid (v/v)
B methanol/glacial acetic acid 8:1 (v/v)
C 10% acetic acid (v/v) and 30% methanol in water
D 1 N NaOH

The sample containing up to 200 µg of protein is filled up to 1.0 ml with ddH_2O. After that, 2.0 ml of Soln. A are added. After mixing, the samples are put into crushed ice for 10 min. Centrifuge the samples thereafter in a refrigerated centrifuge at 4 °C with 4000 × g for 5 min.

Aspirate the supernatant carefully and wash the pellet with Soln. B until the supernatant remains colorless. After the last wash, the precipitate dries at RT.

Dissolve the dry precipitate in 3.0 ml of Soln. D and measure the resulting colored solution in a photometer at 625 nm.

Make the standard curve in the range from 10 to 200 µg BSA.

Modification

Drop samples and standards onto small sheets of glass filter paper (e.g., Whatman GF/A; the sheets are labeled with a pencil). Stain the sheets with Soln. A for 20 min and destain with C until the background is nearly colorless. Extract the sheets with 2.0 ml of Soln. D each after drying and measure the resulting blue solution as described above.

A further modification uses 0.45-µm membrane filters. The procedure is the same as in the first protocol, but instead of centrifugation, the protein-dye complex is sucked through the filters. After that, the filters are washed with Soln. C and extracted with Soln. D. This protocol, given by Nakao et al., is applicable for amounts between 1 and 20 µg.

References

Nakao TM, Nage F (1973) Anal Biochem 55:358
Popov N, Schmitt M, Schulzeck S, Matthies H (1975) Acta Biol Med Germ 34:1441

1.1.5 BCA Protein Determination

This method should be preferred if protein concentration has to be determined in the presence of detergents. But if copper chelators, such as EDTA, or reductants, such as 2-mercaproethanol or DTE/DTT or reducing carbohydrates (e.g. > 10 mM glucose), are components of the sample, the test does not work reliably.

1.1.5.1 BCA Standard Procedure

Solutions/Reagents

A 1% (w/v) BCA (2,2′-biquinoline-4,4′-dicarboxylic acid, bicinchoninic acid, disodium salt), 2% (w/v) $Na_2CO_3 \cdot H_2O$, 0.16% (w/v) disodium tartrate, 0.4% (w/v) NaOH, 0.95% (w/v) $NaHCO_3$, correct pH to 11.25 if necessary with NaOH or $NaHCO_3$. Stable at RT.

B 4% $CuSO_4 \cdot 5H_2O$. Stable at RT.

C mix 100 vol. of Soln. A with 2 vol. of Soln. B. Stable at least for 1 week at RT.

Standard: 1.0 mg/ml BSA[4]

The standard curve is made between 0 and 100 µg.

Fill samples and standards up to 100 µl with ddH_2O, and then add 2.0 ml of Soln. C and incubate the mixture at 37 °C for 30 min. Measuring wavelength is 562 nm.

1.1.5.2 BCA Micromethod

Solutions/Reagents

A 8% (w/v) $Na_2CO_3 \cdot H_2O$, 1.6% (w/v) NaOH, 1.6% (w/v) disodium tartrate, correct pH to 11.25 with $NaHCO_3$. Stable at RT

B 4% (w/v) bicinchoninic acid, disodium salt (BCA). Stable at RT

C 4% $CuSO_4 \cdot 5H_2O$. Stable at RT

D mix 50 vol. of Soln. A with 48 vol. of Soln. B and 2 vol. of Soln. C

The standard curve is made between 0 and 100 µg BSA/100 µl.

Fill up 100 µl of sample or standard to 100 µl with ddH_2O, if necessary, and mix with 100 µl of Soln. D. Read the absorption after incubation for 30 min at 60 °C at 562 nm.

The incubation time may vary dependent on incubation temperature: 37 °C – 2 h.

This microassay may be done on microtiter plates.

References

Fountoulakis M, Juranvill JF, Manneberg M (1992) J Biochem Biophys Meth 24:265

[4] Like by other quantitative methods pure proteins give different results when the same weight was used, e.g., ovalbumin 93%, rabbit IgG 90%, mouse IgG 80%, human IgG 97%, and chymotrypsinogen 100% when compared with BSA (data from Pierce Protein Assay Technical Handbook 1996).

D.A.Harris, C.L.Bashford (ed.) (1987) Spectrophotometry and spectrofluorimetry: a practical approach, IRL Press, Oxford
P.K. Smith, R.I. Krohn, G.T. Hermanson et al. (1985) Anal Biochem 150:76–85

1.1.6 KJELDAHL Protein Determination

Particularly suitable for insoluble proteins, protein in foods and protein covalently immobilized on chromatographic supports.

The KJELDAHL total nitrogen determination method is not very sensitive, but it suits well for analyzing insoluble samples without preceding disintegration. Automated KJELDAHL protein estimations are used especially in food analysis.

Solutions/Reagents

- A selenium reaction mixture for nitrogen determination according to WIENINGER
- B conc. sulfuric acid (98% w/w)
- C 60% NaOH, 10% $Na_2S_2O_3$ (w/v) in ddH_2O
- D 2% boric acid (w/v) in ddH_2O
- E TASHIRO indicator (2 vol. 0.2% methyl red in 90% ethanol + 1 vol. 0.2% methylene blue in 90% ethanol)
- F 0.010 N HCl (standard solution)

Mix 100–250 mg sample exactly weighed with 1.5 g catalyst A, and then add 3 ml of concentrated sulfuric acid B. Heat the mixture at the temperature of boiling sulfuric acid (about 180 °C) for 2 h. Take care that the acid condenses in the middle of the neck of the KJELDAHL flask.

***Caution!** Strongly corrosive! Use a hood.*

Put the flask into the distillation apparatus after cooling; thereafter add slowly 12 ml ddH_2O followed by 12 ml of Soln. C. Heat the mixture to nearly 100 °C and the liberated ammonia is distilled by steam for about 10 min through a condenser, the tip of which is submerged in a flask containing 5 ml of Soln. D. When the distillation is finished (total volume about 25 ml), add three drops of E and titrate the ammonia with Soln. F. The results are calculated as follows:

$$1.0\text{ ml }0.010\text{ N HCl} = 10\ \mu\text{Mol N} = 0.14\text{ mg N}$$

By means of the KJELDAHL factors **F** (Table 1.7) the amount of protein is

$$\text{mg protein} = \text{mg N} \cdot \mathbf{F}$$

and the protein content c of the sample

$$c\ [\%] = \frac{(\text{mg protein}) \cdot 100}{\text{weight of sample [mg]}}$$

Table 1.7. Kjeldahl factors

	Sample	Species	N (%)	Factor **F**
Protein				
	Serum albumin	Man	15.95	6.20
	Serum albumin	Cattle	16.07	6.22
	Ovalbumin	Chicken	15.76	6.34
	Casein	Cattle	15.63	6.39
	Hemoglobin	Pig	16.80	5.95
	Histones	Cattle	18.00	5.55
	Gliadin	Wheat	17.66	5.66
	Legumin	Pea	16.04	5.54
Seed				
	Roe		17.15	5.83
	Wheat		17.15	5.83
	Rice		16.80	5.95
	Beans, peas		16.00	6.25
Animal material	Egg	Chicken	16.00	6.25
	Meat	Cattle	16.00	6.25
	Milk	Cattle	15.67	6.38

References

Jacob S (1965) The determination of nitrogen in biological materials. In: Glick D (ed.) Methods in biochemical analysis vol. 33, p. 241, Wiley, New York

Mazor L (1983) Methods in organic analysis, p. 312, Akadémiai Kiadó, Budapest

1.1.7 UV Photometric Assay of Protein Concentration

The photometric estimation of protein concentration is subject to some special features: Proteins interact with each other depending on their concentration and may change their secondary and/or tertiary structure in a concentration-dependent manner (especially denaturation in diluted solutions). These changes affect the absorption of light, i.e., concentration dependence of molar absorption coefficient ε; therefore, the Beer–Lambert law (eq. e) is not valid over a broad concentration range.

If a compound dissociates in a solvent and one part of the pair has another absorption than the other (e.g., p-nitrophenol/p-nitrophenolate), the absorption coefficient changes with dilution. This should be taken into consideration when different dilutions of a compound are compared. The concentration of an aqueous protein solution can be estimated by reading the UV absorption. The aromatic amino acids (phenylalanine, tryptophan, tyrosine) and the disulfide bond absorb at about 280 nm. The peptide bonds absorb light below 215 nm with much higher absorption coefficients, but in this sector a lot of substances also absorb. Nucleic acids show absorption with a broad band with a maximum around 260 nm. Since for most proteins specific absorption coefficients are unknown and this data are without value in mixtures of several proteins, some equations are developed to overcome this problem. It should be kept in mind that none of these equations gives a "right" value if protein mixtures are measured.

The measurement has to be done against the protein-free solvent (buffer). If a single-beam photometer is used, the absorbance of this blank has to be subtracted from that of the protein solution.

Pay attention to the path of the cuvettes! Equations a) to d) are made for 10.0 mm.

a) WARBURG and CHRISTIAN equation (cf. Protocol 1.2.5):

$$\text{mg protein/ml} = 1.55 \cdot A_{280}^{1\text{cm}} - 0.76 \cdot A_{260}^{1\text{cm}}$$

This estimation of protein concentration is valid up to 20% (w/v) nucleic acid or an A_{280}/A_{260} ratio < 0.6.

b) KALCKAR and SHAFRAN equation:

$$\text{mg protein/ml} = 1.45 \cdot A_{280}^{1\text{cm}} - 0.74 \cdot A_{260}^{1\text{cm}}$$

c) WHITAKER and GRANUM equation:

$$\text{mg Protein/ml} = \left(A_{235}^{1\text{cm}} - A_{280}^{1\text{cm}}\right) : 2.51$$

d) Concentration of immunoglobulins (IgG; cf. Table 8.11 for divergent absorption coefficients)

$$\text{mg IgG/ml} = A_{280}^{1\text{cm}} : 1.38$$

e) BEER–LAMBERT law

$$A_\lambda = \log_{10}\frac{I_0}{I} = \varepsilon_\lambda \cdot c \cdot d$$

A_λ: absorbance at wavelength λ; I_0: intensity of incident light; I: intensity of transmitted light; ε_λ: (molar) absorption coefficient at wavelength λ; c: concentration; d: length of optical path within the cuvette.

If solutions of pure proteins with known amino acid sequence or composition are measured, the concentration c (mol/l) is calculated from the absorbances at 280 nm (A_{280}), 320 nm (A_{320}), 350 nm (A_{350}), and the number of tryptophan (n_{Trp}) and tyrosine residues (n_{Tyr}) and the number of disulfide bridges (n_{S-S}) according to equation f):

f)

$$c = \frac{E_{280} - 10^{(2.5 \cdot \lg E_{320} - 1.5 \cdot E_{350})}}{5540 \cdot n_{Trp} + 1480 \cdot n_{Tyr} + 134 \cdot n_{S-S}}$$

References

Warburg O, Christian W (1941) Biochem Z 310:384
Kalckar HM, Shafran M (1947) J Biol Chem 167:461
Whitaker JR, Granum PR (1980) Anal Biochem 109:155
John RA (1992) In: Eisenthal R, Danson MJ (eds.) Enzyme assays: a practical approach. IRL Press, Oxford, p 59
Welfle H (1996) In: Holtzhauer M (ed.) Methoden in der Proteinanalytik. Springer, Berlin, p 100

1.2 Quantitative Determination of Nucleic Acids

1.2.1 Schmidt and Thannhauser DNA, RNA, and Protein Separation Procedure

This procedure has been developed for quantification of the three types of macromolecules in tissue extracts, where other biomolecules are also present. Small dissolved amounts of DNA, RNA, or protein, especially when no material should be consumed and no interfering substances are in the solution, may be estimated by UV photometry, but a discrimination between DNA and RNA is impossible by reading absorbencies (cf. Protocol 1.2.5).

Solutions/Reagents

A 14% perchloric acid (w/v) in ddH_2O
B 7% perchloric acid (w/v) in ddH_2O
C 3% perchloric acid (w/v) in ddH_2O
D diethylether/ethanol 3:1 (v/v)
E 1 N KOH
F 5% trichloroacetic acid (TCA) (w/v)
G 1 N NaOH

Mix A solubilized, aqueous tissue sample with an equal volume of ice-cold Soln. A, and then add ice-cold Soln. B up to 3.0 ml. The mixture is left for 10 min in an ice bath. After this centrifuge, leave the acidic solution at 0 °C for 10 min with 4000 × g. Wash the pellet three times by resuspension in ice-cold Soln. C and further centrifugation.

Lipids are removed by twofold extraction with ethanol, followed by a threefold extraction with Soln. D at 30–40 °C. Centrifuge the mixtures for a short period at RT between each extraction. Discard the supernatants. After complete lipid extraction the residual pellet is air-dried (Schmidt–Thannhauser powder).

Suspend the dry powder in 0.5 ml of Soln. E at RT. Add 0.5 ml of Soln. A after 1 h and let the mixture for an additional hour in an ice bath. After that, centrifuge the mixture for 10 min at 4000 × g and 0 °C. The supernatant is used for RNA estimation by the orcin method (Protocol 1.2.2): **first supernatant.**

Add 1.0 ml of Soln. F to the above pellet and heat the mixture to 90 °C for 1 h. It is recommended to close the test tubes with a glass ball to avoid dryness.

Important! Avoid loss of solvent!

After cooling, the samples are left for 30 min in an ice bath and are centrifuged again. The supernatants are used for DNA estimation by the diphenylamine method (Protocol 1.2.3): **second supernatant.**

Add 0.5 – 1.0 ml of Soln. G to the precipitate and heat the mixture to 90 °C for 10 min. After cooling, the samples are centrifuged as described above and the supernatant is used for protein estimation by the LOWRY protocol (Protocol 1.1.1.1): **third supernatant.**

References

Schmidt G, Thannhauser SJ (1945) J Biol Chem 161:83

1.2.2 Orcin RNA (Ribose) Determination

Solutions/Reagents

A 15 mg $FeCl_3$ or 25 mg $FeCl_3 \cdot 6H_2O$ in 100 ml conc. HCl
B 100 mg orcin (5-methyl-resorcin) freshly dissolved in 20 ml of Soln. A

Standard 100 µg/ml ribose

1.0 ml of the first supernatant (Protocol 1.1.1) is mixed with 2.0 ml of Soln. B and boiled in a water bath for 30 min. To avoid volume loss, close the test tubes with glass balls.

After cooling to RT, read the absorbance at 660 nm against a blank.

1 µg ribose corresponds to 4.56 µg RNA.

In protein-free and DNA-free solutions the RNA content is determined by UV measurement at 260 nm. 1 O.D. $\approx$ 0.1 µg RNA.

References

Mejbaum W (1939) Z physiol Chem 258:117

1.2.3 Diphenylamine DNA (Deoxyribose) Determination

Solutions/Reagents

A 5% perchloric acid (w/v) in ddH_2O
B 10% perchloric acid (w/v) in ddH_2O
C 300 mg diphenylamine are dissolved in 20.0 ml glacial acetic acid, and then 0.3 ml conc. sulfuric acid are added, followed by 0.1 ml of 50% aqueous acetaldehyde. This reagent has to by prepared freshly.

Standard 100 µg/ml deoxyribose in Soln. A

Mix the SCHMIDT–THANNHAUSER powder with 1.0 ml of Soln. A, and fill up the second supernatant (Protocol 1.1.1) to 1.0 ml with Soln. B.

Heat the acidic solutions to 70 °C for 15 min. After cooling to RT, mix 1.0 ml of the solutions containing 5 – 10 µg deoxyribose (corresponding to 30 – 60 µg DNA) with 2.0 ml of Soln. C, close with Parafilm and incubate overnight at 30 °C.

Read the absorbances of samples and standards at 600 nm.
1 μg deoxyribose = 6 μg DNA.

References

Burton K (1956) Biochem J 62:315

1.2.4 Quantitative DNA Determination with Fluorescent Dyes

For DNA estimations ethidium bromide (EtBr, excitation wavelength λ_{Ex} = 254 or 366 nm, emission wavelength λ_{Em}m = 500–590 nm), 4′,6-diamino-2-phenylindole-2 (DAPI, λ_{Ex} = 372 nm, λ_{Em} = 454 nm), or Hoechst 33 258 (λ_{Ex} = 360 nm, λ_{Em} = 450 nm) are suitable.

Solutions/Reagents

A TE buffer: 10 mM Tris, 1 mM EDTA, made to pH 8.0 with HCl (0.61 g Tris and 0.186 g Na_2-EDTA·$2H_2O$ in 500 ml ddH_2O, aliquoted and autoclaved)
B 1 μg/ml ethidium bromide in TE buffer by dilution of a stock solution of 1 mg/ml EtBr in Soln. A. (The other mentioned dyes are used the same way.)
Standard DNA from calf thymus or phages: 0, 1, 2, 5, 5, 7, 5, 10, and 20 μg DNA per ml TE buffer, stable at −20 °C for several month.

Mix 4 μl of standards and samples with 4 μl of Soln. B. Spot 5 μl of each mixture onto a plastic foil (e.g., Parafilm), lying on a UV transilluminator. Photograph the spots and evaluate the picture densitometrically. The optical densities of the standards are plotted against the amount of their DNA and the DNA content of the samples is calculated from the obtained standard curve (range: 0–15 μg DNA/ml).

The use of PicoGreen for double-stranded DNA (λ_{Ex} = 485 nm, λ_{Em} = 520 nm, linear range 25 pg/ml to 1 μg/ml) and OliGreen (λ_E = 480 nm, λ_{Em} = 520 nm, linear range 100 pg/ml to 1 μg/ml) for oligonucleotides and single-stranded DNA increases the sensitivity up to three orders of magnitude[5]. Using these fluorescent dyes, standards and samples may be measured in a fluorescence microtest plate reader.

References

Daxhelet GA, Coene MM, Hoet PP, Cocito CG (1989) Anal Biochem 179:401
Ausubel FM et al. (eds.) (1994) Current protocols in molecular biology. Wiley, New York, Vol. I, 2.6.7–2.6.8

[5] Haugland RP (1996) Handbook of fluorescent probes and research chemicals, 6th ed. pp. 161–164. Molecular Probes, Eugene, Oregon. Technical Bulletin OliGreen ssDNA reagent, and kit (2005), and PicoGreen reagent and kit (2005), http:// www.invitrogen.com/probes/

1.2.5 Determination of Nucleic Acids by UV Absorption

The UV absorption of nucleic acids depends strongly on their nature and solvent conditions as pH, ionic strength, and temperature. Approximate absorption coefficients are given in Tables 1.8 and 1.9. Using these coefficients, yield is calculated during chromatography, ultracentrifugation, and other preparative processes with sufficient precision.

According to Warburg and Christian the total amount of nucleic acids and protein is calculated using the following equation (the factors F and T are given in Table 1.10 in conjunction with the ratio A_{280}/A_{260}):

$$N = \frac{T \cdot P}{1 - T} = \frac{A_{280} \cdot F \cdot T}{d \cdot (1 - T)}$$

N: concentration of nucleic acids (in mg/ml); P: concentration of protein (in mg/ml); d: optical path (in cm); T: [nucleic acid expressed as part of the total (protein + nucleic acid)]; and factor F from Table 1.10.

References

Warburg O, Christian W (1941) Biochem Z 310:384
Webb JM, Levy HB (1958) In: Glick D (ed.) Methods in biochemical analysis, vol. 6, pp. 1–30, Wiley, New York

Table 1.8. Absorption coefficients of nucleotides (nucleic acids)

	$\varepsilon_{260}^{1\text{cm}} \times 10^{-3}$	$A_{260}^{1\text{mg/ml}}$		
		pH 2	pH 7	pH 12
Ribonucleotides	10.55	10.85	10.3	28.5
Deoxyribonucleotides	10.3	10.45	10.1	22.4

Table 1.9. Conversation factors for nucleic acids

1 $O.D._{260}$ double-stranded DNA (dsDNA)	≅ 50 µg/ml	≅ 0.15 mM nucleotides
1 $O.D._{260}$ single-stranded DNA (ssDNA)	≅ 33 µg/ml	≅ 0.1 mM nucleotides
1 $O.D._{260}$ single-stranded RNA (ssRNA)	≅ 40 µg/ml	≅ 0.11 mM nucleotides
1 µg DNA (1000 bp)	≅ 1.52 pMoles nucleotides	
1 pMol DNA (1000 bp)	≅ 0.66 µg DNA	
1000 kD dsDNA	≅ 1 520 000 bp	
1000 bp DNA	≅ 333 amino acids	≅ 37 000 D protein
50 000 D Protein	≅ 1350 bp DNA	

Table 1.10. The UV quantitation of nucleic acids in the presence of proteins

A_{280}/A_{260}	T	F	A_{280}/A_{260}	T	F
1.75	0	1.118	0.86	0.052	0.671
1.60	0.0030	1.078	0.84	0.056	0.650
1.50	0.0056	1.047	0.82	0.061	0.628
1.40	0.0087	1.011	0.80	0.066	0.605
1.30	0.0126	0.969	0.78	0.071	0.581
1.25	0.0146	0.946	0.76	0.076	0.555
1.20	0.0175	0.921	0.74	0.085	0.528
1.15	0.0205	0.893	0.72	0.093	0.500
1.10	0.0240	0.836	0.70	0.103	0.470
1.05	0.0280	0.831	0.68	0.114	0.438
1.00	0.0330	0.794	0.66	0.128	0.404
0.96	0.0370	0.763	0.64	0.145	0.368
0.92	0.0430	0.728	0.62	0.166	0.330
0.90	0.0460	0.710	0.60	0.192	0.289
0.88	0.0490	0.691			

1.3 Quantitative Phosphate Determinations

1.3.1 Determination of Inorganic Phosphate in Biologic Samples

Solutions/Reagents

A 1 N perchloric acid
B 5 mM sodium molybdate ($Na_2MoO_4 \cdot 2H_2O$, M_r 241.95)

***Important!** Do not use ammonium molybdate!*

C isopropylacetate
Standard 10 mM KH_2PO_4 in 0.5 M perchloric acid

Put 1.5 ml of Soln. B and 2.0 ml of Soln. C into phosphate-free test tubes. The sample, which should contain not more than 100 nmoles phosphate, is mixed with an equal volume of Soln. A. Give 0.5 ml of this mixture to the above mixture of B and C. Shake the obtained mixture vigorously for 30 s and then spin in a centrifuge for a short period to separate the phases. To avoid the decomposition of labile organic phosphates, the extraction should be done at 0 °C or below.

The molybdatophosphate complex remains in the organic phase, which is removed and read at 725 nm against a blank.

Standards are made in the range of 5–100 nmoles phosphate per 0.5 ml.

References

Wahler BE, Wollenberger A (1958) Biochem Z 329:508

1.3.2 Determination of Total Phosphate

> ***Important!*** *Use phosphate-free test tubes (cleaned with hot diluted HCl) or plastic disposables.*

This procedure is simpler and more reliable than that by Fiske and Subbarow.

Solutions/Reagents

A 6 N HCl

B 2.5% ammonium molybdate (w/v) in ddH_2O

C 10% ascorbic acid (w/v) in ddH_2O

D 2% urea (w/v) in ddH_2O

E ***Reagent after ashing***: Soln. B, C, and ddH_2O are mixed in a ratio of 1:1:8

E′ ***Reagent without previous ashing***: Soln. A, B, C, and ddH_2O are mixed in a ratio of 1:1:1:7

E and E′ are stable only for 1 day.

Standard 10 mM KH_2PO_4

conc. sulfuric acid

conc. nitric acid

Ashing

Ashing must be done if phosphate is at least partially covalently bound as, for example, in nucleic acids, nucleotides, or phosphoproteins.

Give 0.2 ml of conc. sulfuric acid to 1–2 ml of aqueous sample. Concentrate the liquid carefully in a hood at about 130 °C and then heat to 280 °C until white fog appears. After cooling, add one to two drops of conc. nitric acid and heat again until nitrous gases are visible. After cooling, add 2 ml of Soln. D, boil the solution for a short period, and fill up to 5.0 ml with ddH_2O.

Determination

Mix 2.0 ml of the sample solution after digestion or sample in Soln. A with Soln. E and E′. Close the test tubes and incubate in the dark at 37 °C for 1.5–2 h. After that, read samples, blank, and standards at 750 nm.

The standard curve is made in the range of 50–350 nmoles phosphate.

100 nmoles phosphate = 9.497 µg PO_4 = 3.097 µg P

References

Chen PS, Toribara TY (1956) Anal Chem 28:1756

1.3.3 Phospholipid Determination

Solutions/Reagents

A chloroform/methanol 1:2 (v/v)
chloroform

Wet the lyophilized sample with 80 µl ddH_2O. After that, add 300 µl of Soln. A and homogenize the sample with a glass-Teflon homogenisator. After addition of 100 µl chloroform, centrifuge the mixture for phase separation. This extraction is repeated three to four times.

Combine the organic phases and vaporize the organic solvents in a nitrogen stream. The remaining solvent is removed in vacuum.

Use the dry residue for digestion and phosphate determination as described in Protocol 1.3.2.

1 nmoles phosphate ≈ 85 ng phospholipid; 1 µg phosphate ≈ 8.3 µg phospholipid with an average M_r of 800.

References

Blight EG, Dyer W (1959) Can J Biochem Physiol 37:911

1.4 Monosaccharide Determination

The quantitative determinations of the monosaccharides ribose and deoxyribose are given in Protocols 1.2.2 and 1.2.3, respectively. The following protocol is useful for all monosaccharides.

Solutions/Reagents

A 80% phenol (w/v) in ddH_2O (phenol must be colorless)
conc. sulfuric acid

Give 1.0 ml of the monosaccharide-containing solution (10–70 µg of saccharide) into a centrifuge tube and mix with 20 µl of Soln. A. Add 2.5 ml of conc. sulfuric acid onto the surface of the liquid (***caution, corrosive!***), then cool for 10 min and keep at 25–30 °C for 10–20 min. After an additional 30 min at RT, read the absorption (hexoses) at 490 nm, pentoses at 480 nm.

The standard curve is made from the appropriate monosaccharide dissolved in water.

> ***Important!*** *Because phenol is used, the waste has to be disposed according to the local regulations.*

Immunochemical determination

Commercially available kits (e.g., Roche Glycan Quantitation Kit) are based on the oxidation of the carbohydrates with periodic acid and subsequent coupling of the formed aldehyde with a hydrazide (e.g., digoxigenin hydrazide). The formed conjugate is estimated immunochemically by ELISA.

References

Dubois M, Gilles KA, Hamilton JK, Rebers PA, Smith F (1956) Anal Biochem 28:350

1.5 Calculations in Quantitative Analysis

The evaluation methodology depends on demand of results: Are only Yes/No answers needed or semiquantitive statements ("strong" – "medium" – "weak"), or are reliable values necessary? To get the first two types of results you have only to choose the right assay protocol and measuring range. For quantitative evaluations often more points of view have to be considered, e.g., rules of statistics. (Some statistic parameters are given in Chap. 9.)

The choice of measuring range is essential for precise results. Most of the quantitative methods have only a relative small linear correlation between measuring signal and amount. As far as this range is covered by the used standards, interpolation between standard and sample is possible. Attention should be paid if extrapolations are used, because especially in the higher range the standard curve becomes flat followed by unacceptable mistakes in calculated values. Figure 1.1 illustrates typical standard curves for protein estimations between 0 and 150 µg (standard: BSA).

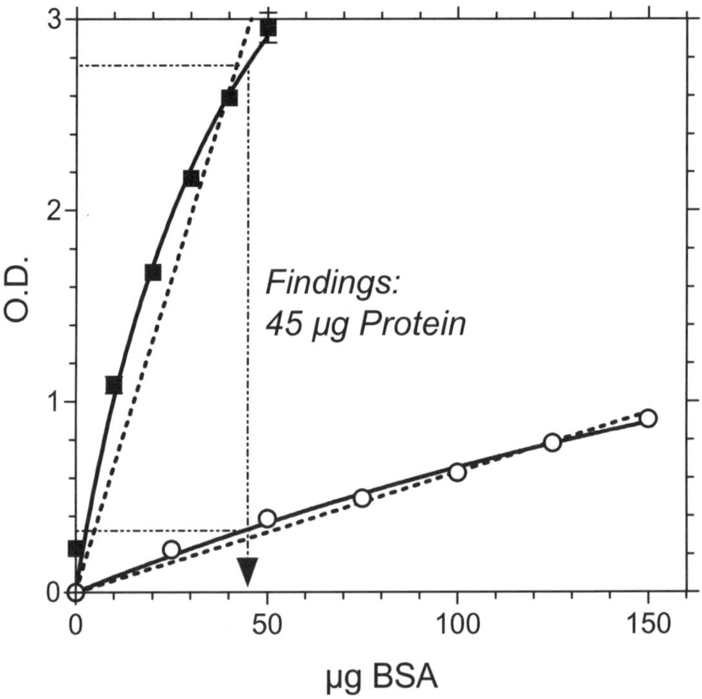

Fig. 1.1. Examples for standard curves resulting from multiple determinations of different amounts of BSA. *Line with circles*: protocol according to LOWRY et al. *Solid line*: nonlinear regression; *dotted line*: linear regressions; wavelength 720 nm. *Line with squares*: BCA protein determination. *Solid line*: nonlinear regression; *dotted line*: linear regression; wavelength 562 nm). *Findings* means an example for graphical evaluation

It should also be kept in mind that the BEER–LAMBERT law often is not valid at higher concentrations, since there occur interactions between chromophores and other molecules[6]. This effect is observed especially at reading of proteins in the UV. The solvent may influence the absorbance too, because, for example, some of the aromatic amino acid residues are buried within hydrophobic core of the molecule and become exposed during unfolding of the protein when the composition of the solvent is changed or the protein is denaturated by dilution.

If you are sure that the BEER–LAMBERT law is fulfilled and a linear correlation between signal and amount is approximately given, the amount of analyte may be calculated by a simple equation:

$$Q_U = \frac{M_U - M_B \cdot Q_S}{M_S - M_B}$$

Q: quantity or concentration; M: analytical signal; U: unknown (sample); S: standard; B: blank.

Controls

Since most of the analytical methods are influenced by an unknown set of circumstances, it is recommended to run each test with appropriate controls. The most important control is the blank, i.e., the assay containing all components with the exception of the analyte. It is also recommended to run each sample in triplicate to get a rational mean and to detect false values.

[6] Galla H-F (1988) Spektroskopische Methoden in der Biochemie. Thieme, Stuttgart

2 Electrophoresis

2.1 Polyacrylamide Gel Electrophoresis Systems

Polyacrylamide gel electrophoresis (PAGE) is the most versatile analytical method in protein analysis. It is a low-cost and reproducible method for comparing and characterizing proteins even from very complex mixtures. It exploits the fact that proteins in aqueous solutions are ions with a positive or negative charge depending on the pH of the environment. During electrophoretic separation ions are driven through a network formed by the hydrophilic gel of cross-linked polyacrylamide. The migration speed depends on the net charge of the molecule, its size, the size of gel mashes, possible interactions between polyacrylamide and macromolecules, and the strength of the electric field.

Other supports in electrophoresis are agarose gels, paper, or celluloses acetate.

Because proteins are amphoteric molecules, they bear negative as well as positive charged residues and their net charge depends on the pH of the used buffer: at pH lager than the isoelectric point (pI) of a given protein negative charges predominate and this protein behaves as an anion; at pH lower than its pI, it is a cation. That means that at a distinct pH some proteins of a sample will move to the anode, whereas others migrate to the cathode. In an electric field the speed of this protein movement depends on the charge and the size of the molecules.

If proteins are in contact with some detergents as sodium dodecylsulfate (SDS) or cetyltrimethylammonium bromide (CTAB) they become denaturated, e.g., their secondary, tertiary, and quaternary structures are destroyed. They get a rod-like shape and the amount of bound detergent is proportionate to the molar mass of the proteins[1]. These protein-detergent complexes have negative charges at slightly alkaline pH if SDS is used and their size (hydrodynamic radius) is approximately proportional to their molar

---

[1] Values of 1.5 mg of SDS per milligram of protein are given (Nielsen TB, Reynolds JA (1978) Meth Enzymol 48:6), but also higher ratios are found (Rao PF, Takagi T (1988) Anal Biochem 174:251: 1.75 to 1.94 mg SDS/protein mg). Glycoproteins often have a much lower SDS/protein ratio (Beeley JG (1985) Glycoprotein and proteoglycan techniques. Elsevier, Amsterdam, p 75).

mass. To check this proportionality a FERGUSON plot should be made (determination of the relative mobility R_f of the protein of interest in dependency of the total concentration of acrylamide%T) or the exact mass is proved by ESI or MALDI mass spectrometry[2].

Relative mobility

The relative mobility of a macromolecule is given by the quotient of its distance of migration measured from the start of separation and the distance of electrophoresis front (position of tracking dye):

$$R_f = \frac{\text{distance}_{\text{macromolecule}}}{\text{distance}_{\text{tracking dye}}}$$

If R_f of different proteins should be compared, the calculations have to be made from the same slab gel to eliminate variances in acrylamide concentration, polymerization, and electrophoretic conditions.

Whether a gel with polyacrylamide concentration gradient or with homogeneous concentration, and whether a denaturating (SDS and/or 2-mercaptoethanol and DTE/DTT[3], or urea) or a non-denaturating (native) system is used, depends on the analytical objective. For a survey or for separation of a mixture of molecules with a broad range of molecular weight, a gradient gel should be chosen, especially because an increased protein sieving effect is observed in the lower molar mass range leading to sharper bands, whereas a homogenous gel should be applied if proteins similar in size and charge will be analyzed. Concerning the geometry of the gel, slab gels are preferred because it is possible to run several samples in parallel under the same conditions.

To illustrate the resolution force of polyacrylamide gels a plot of marker proteins separated in gels with different acrylamide concentrations is given in Fig. 2.1. This figure illustrates that gels with homogenous acrylamide concentration separate over a broad molecular range too, if the migration distance is long enough, but it is observed that a band broadening occurs mostly at higher R_f.

The gel composition is often described by the terms %T and %C. %T refers to the total content of acrylamide (sum of acrylamide and cross-linking monomer), whereas %C is the part of cross-linking substance (e.g., N,N′-methylene bisacrylamide) of monomers.

$$\%T = \frac{AA + C}{V} \cdot 100$$

$$\%C = \frac{C}{AA + C} \cdot 100$$

[2] Discussions of different PAGE systems and possible errors are given by Johnson. (Johnson G (1983) Gel Sieving electrophoresis: a description of procedures and analysis of errors. In: Glick D (ed.) Methods in biochemical analysis, vol 29. Wiley, New York, p 25).

[3] DTE and DTT (CLELAND's reagent) are stereoisomeres, which as well as the optical pure substances or the racemate are effective reductants.

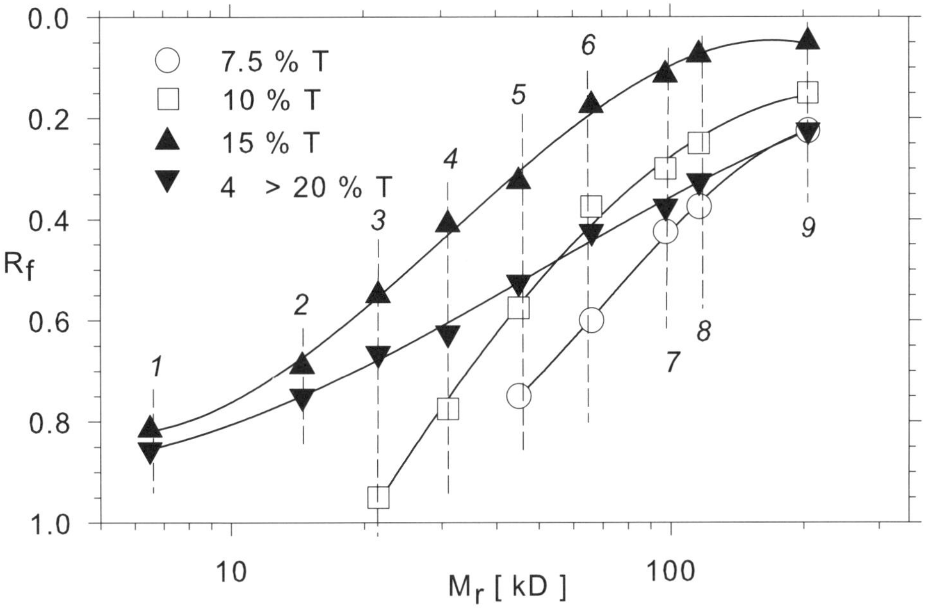

Fig. 2.1. Semilogarithmic plot of molecular weight (M_r) of marker proteins vs relative mobility (R_f) of marker proteins in gels of different acrylamide concentrations %T. Proteins: 1 aprotinin (6.5 kD); 2 lysozyme (14.5 kD); 3 soybean trypsin inhibitor (21.5 kD); 4 carbonic acid anhydrase (31 kD); 5 hen ovalbumin (45 kD); 6 bovine serum albumin (66 kD); 7 phosphorylase b (97.4 kD); 8 β-galactosidase (116 kD); 9 myosin (205 kD)

AA: amount of acrylamide (g); C: amount of cross-linker (g); V: volume of gel mixture (ml).

Self-made PAGE gels should have at least 50 mm separation distance and a width of 0.75–1 mm. Generally, for analytical purposes slab gels of 60 × 80 × 1 mm are sufficient, but also larger gels are usable, especially for semi-preparative purposes.

The electrophoresis is controlled by a power supply mostly in the constant current (cc) or constant voltage (cv) mode. It is advised against application of constant current during separation, because during the run the electric resistance increases, and because current is constant, voltage and heat production increase, too.

Salt effects in PAGE

Most of the electrophoresis systems are sensitive to ionic strength ("salt content") of samples. Dialysis of the sample against a 100-fold volume of sample buffer for 30 min or protein precipitation by TCA and dissolving of the pellet in sample buffer obviate this problem.

PAGE systems described in this chapter are well established; nevertheless, modifications concerning acrylamide concentration (%T as well as %C) may be done to optimize the separation conditions.

***Caution!** Acrylamide is very toxic. Avoid inhalation of dust and wear gloves during manipulation with monomers!*

References

Allen RC, Saravis CA, Maurer HR (1984) Gel electrophoresis and isoelectric focusing of proteins. Selected techniques. W. de Gruyter, Berlin
Chrambach A, Dunn MJ, Radola BJ (eds.) (1987 foll.) Advances in electrophoresis, vol. 1 foll., VCH, Weinheim
Hames BD, Rickwood D (eds.) (1990) Gel electrophoresis of proteins: a practical approach, 2nd ed., Oxford University Press, New York
Deyl Z, Chrambach A, Everaerts FM, Prusik Z (eds.) (1983) Electrophoresis. A survey of techniques and applications. J Chromatogr Library, vol. 18B, Elsevier, Amsterdam
Westermeier R (1997) Electrophoresis in practice: guide to methods and applications of DNA and protein separations. 2nd ed., VCH, Weinheim

2.1.1 Laemmli SDS-Polyacrylamide Gel Electrophoresis

The Laemmli SDS-PAGE protocol is one of the most important analytical techniques in analytical protein separation. It is a system with discontinuous pH gradient (disc electrophoresis) and consists of a stacking and a separation gel different in acrylamide concentration and pH. The separation gel may be formed with homogenous acrylamide concentration or with an increasing gradient.

Molar mass determinations based on SDS-PAGE is sometimes misleading, since some proteins are not conversed completely into a rod-like shape or the protein/SDS ratio differs from the average.

Solutions/Reagents

- A 40.0% acrylamide (w/v), 1.08% N,N′-methylene bisacrylamide (w/v) in ddH_2O; (%C = 2.63)
- A′ 40.0% acrylamide (w/v), 2.1% N,N′-methylene bisacrylamide (w/v) in ddH_2O; (%C = 5.0)
- B 1 M Tris · HCl, pH 8.8
- C 0.5 M Tris · HCl, pH 6.8[4]
- D 20% SDS (w/v) in ddH_2O (store at RT, below 10 °C precipitation occurs)
- E 10% TEMED (v/v) in ddH_2O
- F 10% ammonium persulfate (APS) (w/v) in ddH_2O (stabilize for 1 day at 2–8 °C in a closed tube)
- G *sample buffer (reducing buffer)*: 50 mM Tris · HCl, pH 6.8, 4% SDS (w/v), 10 mM DTE[5] (v/v), 10 mM Na_2-EDTA

[4] Addition of 0.01% bromophenol blue to buffer C makes it easier to recognize of the slots of the stacking gel and also allows the identification of the electrophoresis front during run.

[5] DTE/DTT (Cleland's ragent) as reducing reagent is recommended instead of 2-mercaptoethanol, because the latter often forms pseudobands especially in silver-stained gels. Forming of new disulfide bonds by air or peroxidisulfate is suppressed by addition of 10 mM (final concentration) N-methylmaleiimide (NEM) or iodoacetamide. A further selcective reduction reagent is tris(2-carboxethyl)-phosphonium chloride (TCEP · HCl), 100 mM stock solution in H_2O, final concentration

or

G′ *sample buffer (reducing buffer)*: 50 mM Tris · HCl, pH 6.8, 10% glycerol (v/v) or 10% sucrose (w/v), 4% SDS (w/v), 10 mM DTE, 10 mM Na_2-EDTA, 0.005% bromophenol blue (w/v), 0.005% Pyronin Y[6]

or

G″ *sample buffer 4fold (nonreducing buffer)*: 200 mM Tris · HCl, pH 6.8, 20% glycerol (v/v) or 20% sucrose (w/v), 16% SDS (w/v), 40 mM Na_2-EDTA, 0.02% bromophenol blue (w/v), 0,02% Pyronin Y (w/v)[7]

H *electrode buffer*: 50 mM Tris (6.0 g/l), 384 mM glycine (28.8 g/l), 0.1% SDS (1 g/l), pH 8.3

Soln. A and A′ should be made from substances of highest available purity and amount should be not more than a monthly consumption. Freeze the stock solutions in aliquots until use. The solutions must not be heated above 30 °C to avoid the formation of acrylic acid. Addition of a small amount of mixed-bed ion exchanger catches traces of acrylic acid and depresses interactions between gel matrix and protein.

Preparation of slab gels

The gel cassette is assembled according to manufacturer's instructions. Make sure that the glass plates are clean. If necessary, clean the plates with a common washing liquid, rinse thoroughly with water, and dry. Use plain spacers. Sealing of the cassette with self-adhesive tape, molten 1% agarose, or a small amount of polymerization mixture should be done only in exceptional cases.

To allow a complete polymerization the separation gel should be made the day before use. To get a smooth surface and to avoid drying the gel, cover the liquid polymerization mixture with a layer of ddH_2O, 1:20 diluted buffer B, or n-butanol. When water or buffer is used, add these liquids at both sides of the sandwich and allow to come together slowly in the middle. After completing the polymerization reaction, a sharp borderline appears between polyacrylamide gel and liquid.

The stacking gel is made just before performing the electrophoresis. The residual liquid on top on the separation gel is removed completely using small pieces of filter paper. Then the stacking gel mixture is added and the slot former ("comb") is in-

---

20 mM; but it decompose fast: after 20 h at pH 7.0 remain only 30% of active reagent.

[6] Pyronin Y (Pyronin G, C.I. 45005) moves some faster than bromophenol blue and thus better indicates the electrophresis front. A further benefit is the strong binding to nitrocellulose which allows the identification of the electrophoresis front after immunochemicl reactions in Western blotting.

[7] If a comparison of reduced and nonreduced samples is intended, 1/10 sample volume of 0.1 M DTT in H_2O (stable for month if frozen) will be added to an aliquote prior to heating.

serted. Be sure that no air bubbles are between or under the teeth. The distance between the lower side of teeth and the borderline between stacking and separation gel should be at least 2 mm and not more than 5 mm.

Pore gradient gels

Acrylamide concentration (pore) gradient gels are made using a gradient mixer as shown in Fig. 2.2. The chamber at the outlet is filled with the required amount of higher concentrated (dense) mixture (ds), and then the connection between the chambers is opened for a moment to remove air bubbles (compositions of acrylamide mixtures are given in Table 2.2). When the lower concentrated mixture is poured into the respective chamber, the stirrer is switched on and the needed amount of Soln. F is added by intensive, but careful, mixing. Immediately after starting the polymerization by addition of ammonium persulfate, open both plugs and pour the acrylamide mixture into the gel cassette. Note that the volume of dense and light mixture must be exactly the desired volume of separation gel.

When the solutions are completely run out, the liquid in the cassette is overlaid as described above. To stabilize the concentration gradient and to delay polymerization, the cassette should be put carefully into a refrigerator.

Protocols for polyacrylamide gels with different concentrations are given in Tables 2.1 and 2.2.

After polymerization, just before performing the electrophoresis, the stacking gel is prepared as described previously.

The gel cassette is mounted into the electrophoresis apparatus and electrophoresis tanks are filled with electrode buffer (Soln. H). Remove air bubbles at the interfaces between electrode buffer and

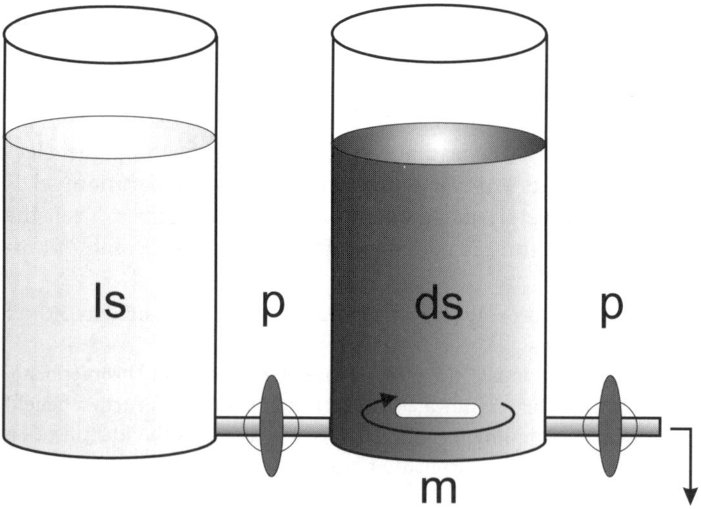

Fig. 2.2. Gradient mixer. *ls* light solution, *ds* dense solution, *m* magnetic stirrer, *p* plug valve

Table 2.1. Recipes for casting gels with homogeneous acrylamide concentration (LAEMMLI system)

Solution	%T = 5.0[a]	7.5[a]	10.0[a]	12.0[b]	15.0[b]	20.0[b]	Stacking gel[c]
				(ml/10 ml)			
A	1.22	1.83	2.43	–	–	–	1.10
A′	–	–	–	2.97	3.55	4.75	–
B	3.75	3.75	3.75	3.75	3.75	3.75	–
C	–	–	–	–	–	–	2.50
D	0.05	0.05	0.05	0.05	0.05	0.05	0.05
E	0.10	0.10	0.10	0.10	0.10	0.10	0.10
			Add H_2O to 9.90 ml				
F	0.10	0.10	0.10	0.10	0.10	0.10	0.10

[a] %C = 2.68
[b] %C = 5.0
[c] %T = 4.5, %C = 2.68

Table 2.2. Recipes for casting gels with acrylamide concentration gradient 5 → 15 and 5 → 20%, respectively (LAEMMLI system)

Solution	Light solution %T = 5.0% %C = 2.68	Dense solution %T = 15.0	%T = 20.0
		%C = 5	
	(ml/10 ml; total volume 20 ml)		
A	1.22	–	–
A′	–	3.55	4.75
B	3.75	3.75	3.75
D	0.05	0.05	0.05
E	0.05	0.05	0.05
	H_2O to 9.95	Add glycerol to 9.95 ml	
F	0.05	0.05	0.05

[a] For gradients with other %T and %C, amounts of Soln. A and A′, respectively, are given in Table 2.1

gel. The electrode buffer may be used for several times if pH is not shifted and if it is not contaminated by constituents of the samples.

> ***Important!*** *Remove the comb after filling the electrophoresis chambers with electrode buffer and remove air bubbles within the slots.*

Sample preparation

If the sample buffer G is used for dissolving precipitated or lyophilized samples (sometimes such samples are difficult to dissolve in Soln. G′), some glucose, sucrose, or glycerol and up to 5% of sample volume of saturated bromophenol blue in Soln. G may be added to ease sample application.

If potassium ions are present within the sample, potassium dodecylsulfate precipitates and dodecylsulfate ion concentration is decreased dramatically; therefore, potassium should be changed to sodium by dialysis or gel filtration of the sample.

Mix liquid samples with sample buffer Soln. G and G′ in a 1:1 ratio or 3:1 with Soln G″. Then heat the samples to 95 °C for 5 min and centrifuge to collect condensed water. If the samples are stored frozen, heat them again prior to application since the formed SDS/protein complex is sometimes unstable.

The sample must have a blue color. If not, adjust pH with some microliters of Soln. B.

If unknown samples have to be analyzed by Western blotting, it is recommended to run the sample as well as reduced and nonreduced, and heated and unheated, since sometimes epitopes are destroyed when disulfide bridges are cleaved or reconstitution fails after withdrawing of SDS.

To avoid reduction of an unreduced sample by its neighbor during electrophoresis, inactivate the DTE overload of the reduced sample by iodoacetamide or NEM.

The applied sample volume depends on the slot size and should be as low as possible, and the concentration should be as high as possible. As an example, with 1 mm gel thickness and 10 to 15 cm separation gel length, the amount of protein should not exceed 20–30 µg/mm^2 slot bottom area.

The minimal amount of applicable protein depends on the detection method. Mostly detection by Coomassie staining is tenfold less sensitive than by silver staining, which drops down to 1 ng per band.

Add the samples with a syringe (e.g., Hamilton syringe) or special pipette tips to form a lower layer below the electrophoresis buffer within the slots.

Start the electrophoresis immediately when all samples, marker protein mixtures, or references are applied, because molecules diffuse through the soft stacking gel and the pH jump between stacking gel and separation gel, which is important for separation power, drops down.

Electrophoresis

Connect the electrophoresis apparatus to the power supply and switch on the voltage (time course) directly after sample application.

Check the proper connection by wires: In SDS electrophoresis the migration is "–" (mostly black wire) → "+" (mostly red wire), when cetyltrimethylammonium bromide (CTAB) is used instead of SDS the direction is "+" → "–".

Voltage during electrophoresis

As a general rule, a voltage of 30–40 volts is useful for introducing the sample into the stacking gel and field strength of 10–15 V/cm separation gel length for the separation is sufficient. If an efficient cooled device is used, the field strength may be increased.

***Caution!** Direct current used in electrophoresis is very dangerous. Never disconnect electrodes from the electrophoresis chambers before switching off the power supply!*

> ***Important!*** *Cool the whole polyacrylamide gel during electrophoresis, since most of the electric power produces* COULOMB *heat resulting in "smiling" bands.*

The electrophoresis is finished when the tracking dye has reached about 5 mm above the bottom of the gel. Turn off the power supply, open the cassette carefully, and submerse the gel into a tenfold of gel volume of fixation solution [15% trichloroacetic acid (w/v), 15% 5-sulfosalicylic acid (w/v), or glacial acetic acid/methanol/H_2O (1:3:6, v/v/v)] or mount the gel into the electrotransfer apparatus. Protein fixation

If necessary, store the gel within the cassette or wrapped with plastic film in a refrigerator overnight, but avoid freezing.

References

Laemmli UK (1970) Nature 227:680

2.1.2 SDS-Polyacrylamide Gel Electrophoresis at Neutral pH (NuPAGE)

Some proteins with posttranslational modifications are sensitive even with respect to the slight alkaline pH of the LAEMMLI system. For example, phosphate groups of phosphoproteins may hydrolyze or SH groups are possibly oxidized. To prevent this, the neutral pH SDS-PAGE system is used. Further advantages are shorter separation time by using higher voltages and a higher separation power. NuPAGE[8] gels and buffers are available ready to use, but they are also easily self-made.

Solutions/Reagents

- A 39% acrylamide (w/v), 1% N,N′-methylene bisacrylamide (w/v)
- B 2.86 M Bis-Tris (M_r 209.24), 1.71 M HCl (3.84 ml 37% HCl per 25 ml), adjusted to pH 6.5
- C 10% TEMED (v/v) in ddH_2O (stabile at RT)
- D 10% ammonium persulfate (w/v) in ddH_2O (stabilize for 2–3 days at 4 °C)
- E ***MES electrophoresis buffer 20-fold***: 1 M MES (M_r 195.2), 1 M Tris, 2% SDS (w/v), 0.6% EDTA (free acid) (w/v), pH 7.3
- F ***MOPS electrophoresis buffer 20-fold***: 1 M MOPS (M_r 209.3), 1 M Tris, 2% SDS (w/v), 0.6% EDTA (free acid) (w/v), pH 7.7
- G ***sample buffer fourfold***: 0.986 M Tris, 40% sucrose (w/v), 8% SDS (w/v), 0.06% EDTA (w/v), 0.075% (w/v) Coomassie Brilliant Blue R250, 0.25%) phenol red (w/v)
- H ***blotting buffer 20-fold***: 0.5 M Bicine (N,N-bis(2-hydroxyethyl)-glycine), 0.5 M Bis-Tris, 20 mM EDTA

Mix the gel according to Table 2.3 and pour it into the cassette. Since a stacking gel is not necessary, put the comb directly into

[8] Invitrogen Corp./NOVEX

Table 2.3. Casting protocol for neutral-pH SDS-PAGE (%C = 2.5)

Solution	%T = 4	8	10	12
		(ml/10 ml)		
A	1.00	2.00	2.50	3.00
B	1.25	1.25	1.25	1.25
C	0.02	0.02	0.02	0.02
		Add H_2O to 9.98 ml		
D	0.02	0.02	0.02	0.02

the separation gel. After complete polymerization, the gel is ready for use.

Mount the gel into the electrophoresis apparatus; fill the electrode chambers with 1:20 diluted MES or MOPS electrode buffer.

A concentration gradient gel with 4 → 12%T running with MES buffer separates between 2 and 250 kD; performing with MOPS buffer the separation range is between 10 and 250 kD.

Mix samples with one-third of their volume of buffer G and heat at about 70 °C for 5 min. For cleavage of disulfide bridges add DTT to a final concentration of 10 mM.

Run a gel of 8 cm separation length with 200 V cv for 50–60 min. Surround the gel completely by electrode buffer during electrophoresis to carry off the heat.

After electrophoresis, the gel is fixed and stained as other PAGE gels, too. For semi-dry blotting the 1:20 diluted buffer H is recommended.

References

NuPAGE Electrophoresis System. Instruction Booklet (1997) NOVEX, San Diego, Calif.

2.1.3 SDS-Polyacrylamide Gel Electrophoresis According to Weber, Pringle, and Osborn

This PAGE is a continuous system with respect to pH. Its resolution is lower than that of a disc system. The advantage lies in the use of buffers free of primary amino groups; therefore, it is recommended if an electrotransfer is intended onto chemical reactive supports because a buffer change decreases transfer yield and separation performance (broadening of bands by diffusion during buffer change).

Solutions/Reagents

A 44.4% acrylamide (w/v), 1.2% N,N′-methylene bisacrylamide (w/v) in ddH_2O (%C = 2.65)

B 0.2 M sodium phosphate buffer, pH 7.2, 0.2% SDS (w/v) (**do not use potassium phosphate**) (7.8 g $NaH_2PO_4 \cdot H_2O$, 38.6 g $Na_2HPO_4 \cdot 7H_2O$ and 2 g SDS in 1000 ml ddH_2O)

C TEMED (v/v) in ddH_2O (stable at RT)

D 10% ammonium persulfate (w/v) in ddH_2O (stable for 2–3 days at 4 °C)

E ***sample buffer***: 10 mM sodium phosphate buffer, pH 7.2, 2% SDS (w/v), 4% 2-mercaptoethanol (v/v)

For preparing the separation gel, pipette the solutions in order given in Table 2.4, adjust the required volume with ddH_2O, and mix well (avoid foam forming). Start the polymerization reaction by addition of Soln. D. Introduce the mixture into the cassette without air bubbles up to about 10 mm from the top of the front plate and gently layer ddH_2O or n-butanol to keep a smooth surface. Polymerization needs 20–30 min and may be prolonged by cooling or decreasing the amount of Soln. D, and vice versa.

Just before performing the electrophoresis, carefully remove the liquid above the gel with filter paper and prepare the stacking gel. Since it is a continuous separation system, the stacking gel may be poured immediately after introducing the separation gel instead of covering with ddH_2O or n-butanol, but getting sharp separations between both gels is a somewhat sophisticated process.

Prepare the sample in buffer E with a protein concentration not more than 20 mg/ml. Heat the solution to 95 °C for 2–3 min and supplement with some crystals of sucrose or a droplet of glycerol or 1/10 volume of bromophenol blue in 50% sucrose solution.

Mix the dissolved sample with an equal volume of double concentrated sample buffer E.

If the sample has an (expected) ionic strength > 0.05 it has to dialyze against sample buffer because this system is exceptionally sensitive to ions.

The minimal amount of loaded protein depends on the detection method. As a rule of thumb, for Coomassie staining at least 1 µg of protein per band is needed.

Table 2.4. Casting protocol for the WEBER, PRINGLE, and OSBORN system (C = 2.65%)

Solution	%T = 20	15	10 (ml/10 ml)	7.5	5	Stacking gel
A	4.40	3.30	2.20	1.64	1.10	0.72
B	5.00	5.00	5.00	5.00	5.00	5.00
C	0.10	0.10	0.10	0.10	0.10	0.10
			Add H_2O to 9.9 ml			
D	0.10	0.10	0.10	0.10	0.10	0.10

Electrode buffer is Soln. B, diluted 1:1 with ddH_2O. Voltage regime and fixation are the same as for the LAEMMLI system (Protocol 2.1.1), staining as usual.

References

Weber K, Pringle JR, Osborn M (1972) Meth Enzymol 26:3

2.1.4 Urea-SDS-Polyacrylamide Gel Electrophoresis for the Separation of Low Molecular Weight Proteins

The presence of urea as well as the high concentration of acrylamide (%T = 13.35, %C = 6.22) allow the separation especially of low molecular weight proteins and polypeptides.

Solutions/Reagents

- A 0.6% TEMED (v/v), 0.8% SDS (w/v), 0.8 M phosphoric acid, adjusted with solid Tris to pH 6.8
- B 37.5% acrylamide (w/v), 1.25% N,N′-methylene bisacrylamide (w/v) in ddH_2O
- C 6.85 M urea, 50 mM phosphoric acid, 2% SDS (w/v), 1% 2-mercaptoethanol (v/v), 0.1 mM EDTA, adjusted with solid Tris to pH 6.8
- D ***electrode buffer***: 0.1 M phosphoric acid, 0.1% SDS (w/v), adjusted with solid Tris to pH 6.8

Before pouring the gel, solution A′ is made by dissolving 1 mg of ammonium persulfate in 1 ml Soln. A. Solution B′ is also made freshly by dissolving of 4.114 g urea and 0.125 g N,N′-methylene bisacrylamide in 10 ml of Soln. B (do not heat above 30 °C).

Mix the gel gently according to Table 2.5 and introduce it into the cassette without delay. Insert the comb into the liquid polyacrylamide mixture. Polymerization proceeds at room temperature.

Samples are mixed with an equal volume of Soln. C.

Fill the electrophoresis chambers with electrode buffer D and perform separation with 6–8 V/cm (with cooling) or 3 V/cm at RT overnight. It is not recommended to use the whole gel length for separation, but to stop the run if the tracking dye has moved three quarters of the gel.

Table 2.5. Casting protocol of urea-SDS-PAGE

Solution	ml/10 ml
A′	1.25
B′	4.70
C	4.05

Solutions A′ and B′ (see text)

Fix the gel thoroughly either in TCA or 5-sulfosalicylic acid (cf. Protocol 2.1.1) after electrophoresis to remove urea prior to staining.

References

Swank RT, Munkres KD (1971) Anal Biochem 39:462

2.1.5 TRICINE-SDS-Polyacrylamide Gel Electrophoresis for Proteins and Oligopeptides in the Range of 1000–50 000 Daltons

The SCHÄGGER and VON JAGOW system is also suitable for small proteins and polypeptides, especially for peptides resulting from mapping experiments after bromocyan or tryptic cleavages. Because of its excellent separation power in the lower molecular weight range, it is a good completion to the SWANK and MUNKRES protocol (Protocol 2.1.4). Which of these systems is used depends on the analytical aim.

Solutions/Reagents

A 48% acrylamide (w/v), 1.5% N,N′-methylene bisacrylamide (w/v)
A′ 46.5% acrylamide (w/v), 3.0% N,N′-methylene bisacrylamide (w/v)
B 3 M Tris, adjusted with HCl to pH 8.45, 0.3% SDS (w/v)
C 10% TEMED (v/v) in ddH_2O
D 10% ammonium persulfate (w/v)
E ***anode buffer***: 0.2 M Tris-HCl, pH 8.9
F ***cathode buffer***: 0,1 M Tris, 0,1 M Tricine (M_r 179.2), pH 8.25, 0.1% SDS (w/v)
G ***sample buffer fourfold***: 150 mM Tris, 12% SDS (w/w), 50 mM DTE or DTT, 30% glycerol (v/v), 0.05% (w/v) Coomassie Brilliant Blue G250, pH 7.0

Prepare the separation gel and the stacking gel according to Table 2.6. The separation gel may be poured separately and overlaid with ddH_2O, but the stacking gel also can be carefully poured directly onto the fresh (liquid) separation gel.

For gels with %T > 10 and %C > 3 it is recommended to use a separation gel between stacking and separation gel (about 1/5 to 1/10 of the total gel length).

Dissolve the samples in buffer G (0.5–2 µg protein or 2–5 µg peptide per expected band, depending on the detection method). Mix liquid samples in a ratio of 3 vol. of sample to 1 vol. of buffer (fourfold), dissolve solid samples in 1:4 diluted buffer G, heat to 80 °C[9] for 10 min, and add the samples to form a layer below the cathode buffer in the sample pockets of the gel.

[9] If urea is added to the sample buffer, heat only to 40 °C for 20 min.

Table 2.6. Casting protocol for TRICINE-SDS-PAGE. (According to SCHÄGGER and VON JAGOW)

Solution	Stacking gel[a]	Spacer gel	Separation gel T% = 10 C% = 3 (ml)	%T = 16.6 %C = 3	%T = 16.5 %C = 6[b]
A	0.80	2.00	2.00	3.33	–
A′	–	–	–	–	3.33
B	2.5[b]	3.33	3.33	3.33	3.33
C	0.075	0.05	0.05	0.033	0.033
Glycerol	–	–	1.0	1.0	1.0
			Add H_2O to 10 ml		
D	0.10	0.05	0.05	0.033	0.033
			(V/cm gel length)		
cv during run			8–11	6–7	6.5–7.5

[a] Addition of a trace of Coomassie Brilliant Blue R250 facilitates the sample application and indicates the electrophoresis front
[b] For peptides < 5 kD

As tracking dye Coomassie Brilliant Blue G-250 is recommended. Bromophenol blue does not suit well since it moves significantly slower than small peptides.

Run the electrophoresis at room temperature as indicated in Table 2.6.

After the run common fixation and staining protocols are used.

For Western blotting according to the KYHSE-ANDERSON semidry protocol (Protocol 2.4.3), the following buffers are recommended by SCHÄGGER and v. JAGOW:

cathode buffer: 0.3 M ε-aminohexanoic acid, 30 mM Tris, pH 8.6;

anode buffer: 0.3 M Tris, 0.1 M TRICIN, pH 8.7; it is possible to supplement with 20% methanol (v/v).

References

Schägger H, v. Jagow G (1987) Anal Biochem 166:368
Jagow G von, Schägger H (1994) A practical guide to membrane protein purification. Academic Press, San Diego, p 65

2.1.6 SDS-Polyacrylamide Gel Electrophoresis at pH 2.4

This system allows the separation of alkali-labile proteins (e.g., acylphosphate phosphoproteins) under denaturating conditions according to their molar mass. Despite the low acrylamide concentration (%T = 5.61, %C = 3.61), the separation force is remarkable. Because it is a SDS-containing system, the migration is from "–" to "+" despite the low pH.

Solutions/Reagents

A 40.0% acrylamide (w/v), 1.5% N,N′-methylene bisacrylamide in ddH_2O

B 1 M sodium phosphate-phosphoric acid buffer, pH 2.4

C 20% SDS (w/v) in ddH_2O

D 5 mM ascorbic acid

E 0.025% $Fe(II)SO_4 \cdot 7H_2O$ (w/v) in ddH_2O

F 2.5% H_2O_2

G ***sample buffer***: 50 mM sodium phosphate buffer, pH 2.4, 2% SDS (w/v), 2% 2-mercaptoethanol (v/v) or DTT

H ***electrode buffer***: 50 mM sodium phosphate buffer, pH 2.4, 0.1% SDS

Mix the gel according to Table 2.7, pour the mixture into the cassette and cover with Soln. I (Table 2.7). The polymerization takes place very slowly; therefore, the gel should be prepared 36–48 h before electrophoresis. A stacking gel is not needed.

Dissolve samples in buffer G. Electrode buffer is Soln H. Applied voltage should be 8–10 V per centimeter of gel length. Tracking dye has to be an anionic one intensively colored at pH 2.4 (e.g., cresol red or Pyronin Y).

References

Auruck J, Fairbanks G (1972) Proc Natl Acad Sci USA 69:1216

Table 2.7. Casting protocol for the acidic SDS-PAGE

Solution	Gel (ml/10 ml)	Buffer I
A	1.40	–
B	0.50	–
C	0.50	0.05
D	1.00	1.00
E	0.10	0.10
	Add H_2O to 9.99 ml	
F	0.01	0.01

2.1.7 Urea-Polyacrylamide Gel Electrophoresis for Basic Proteins at pH 2

The SDS-free PAGE system described by Panyim and Chalkley is suitable especially for small basic proteins (e.g., histones). Additives such as non-ionic detergents as Triton X100 [0.05–0.1% (w/v)] may additionally increase the resolution.

Solutions/Reagents

A 60.0% acrylamide (w/v), 0.4% N,N′-methylene bisacrylamide (w/v) in ddH_2O

B 14.5% acetic acid (w/v), 5.4% TEMED (v/v) in ddH_2O
C 4 M urea in ddH_2O
D ***electrode buffer***: 2.5 M urea, 0.9 M acetic acid

Acrylamide and urea must be ultra-pure. Their aqueous solutions are not allowed to warm above 35 °C.

The gel resulting from Table 2.8 has %T = 15.1 and %C = 0.66. A stacking gel is not required.

Prior to protein separation, a pre-electrophoresis is essential. It is done overnight with 3.5 mA/cm^2 (constant current). The electrode buffer D is renewed after pre-electrophoresis.

The samples are dissolved in buffers C or D, supplemented with some crystals of sucrose.

The electrophoresis is run at 8–10 V/cm (cv), the proteins migrate from "–" to "+". The electrophoresis front may be indicated by neutral red.

References

Panyim S, Chalkley R (1969) Biochem Biophys Res Comm 37:1042

Table 2.8. Casting protocol for the acidic urea PAGE

Solution	(ml/10 ml)
A	2.50
B	1.25
C	6.25
Ammonium persulfate (added as solid)	12.5 mg

2.1.8 Anodic Discontinuous Polyacrylamide Gel Electrophoresis (Native PAGE)

If proteins with acidic or nearly neutral pI shall be electrophoretically analyzed without destroying (denaturating) biological activity, this PAGE system may be used. In contrast to SDS-containing systems, which sieve according to molecular size, the separation principle is electrophoretic mobility, i.e., molecular size ("molar mass") is not the primary quantity; netto charge at the given pH is more dominant.

Quaternary structures or ligand-ligate (receptor) interactions may be partially conserved during electrophoresis. Identification of a distinct protein is possible only by biochemical or immunochemical reactions or by comparison with an authentic sample.

Most proteins migrate from cathode ("–") to anode ("+"). As tracking dye for indicating the electrophoresis front may serve bromophenol blue.

Solutions/Reagents

A 18.25 g acrylamide, 0.487 g N,N′-methylene bisacrylamide in 50 ml ddH_2O
B 0.75 M Tris · HCl, pH 8.9
C 0.125 M Tris · HCl, pH 6.8
D 0.05 M Tris, 0.38 M glycine, adjusted with HCl to pH 8.3
E 10% TEMED (v/v) in ddH_2O
F 10% ammonium persulfate (w/v) in ddH_2O

Solutions are gently mixed according to Table 2.9 in the indicated order and filled up with ddH_2O. Start the polymerization by addition of Soln. F and pour the mixture immediately into the gel cassette. When the appropriate height is reached, cover the liquid with water or n-butanol to get a smooth surface. Prepare the stacking gel as short as possible before starting the electrophoresis to avoid a decrease of the pH jump between stacking and separation gel by diffusion.

Electrode buffer is Soln. D, the sample buffer is prepared by addition of a trace bromophenol blue and sucrose to Soln. D. Electrophoretic conditions are similar to that given in Protocol 2.1.1.

References

Davis BJ (1964) Ann NY Acad Sci 121:404

Table 2.9. Casting protocol for native PAGE

Solution	Stacking gel[a] (ml/10 ml)	Separation gel[b] (ml/10 ml)
A	1.2	2.0
B	–	5.0
C	5.0	–
E	0.1	0.1
	Add H_2O to 9.9 ml	
F	0.1	0.1

[a] %T = 4.5, %C = 2.6
[b] %T = 7.5, %C = 2.6

2.1.9 Cathodic Discontinuous Polyacrylamide Gel Electrophoresis (Native PAGE)

This protocol is adapted for analysis of proteins with pI > 8. The other deviations from SDS-PAGE are described in Protocol 2.1.8.

> ***Attention!** Migration is from "+" to "–", i.e., the opposite to SDS-PAGE or anodic systems.*

Basic fuchsin (rosaniline, basic violet 14) or Pyronin Y are suitable as tracking dye.

Solutions/Reagents

A 18.25 g acrylamide, 0.487 g methylene-bisacrylamide in 50 ml ddH_2O
B 0.12 M KOH, 0.75 M acetic acid, pH 4.3
C 0.12 M KOH, 0.125 M acetic acid, pH 6.8
D 0.35 M β-alanine, 0.14 M acetic acid, pH 4.5
E 10% TEMED (v/v) in ddH_2O
F 10% ammonium persulfate (w/v) in ddH_2O

The pipetting scheme is given in Table 2.9. Hints for preparing the gel are described in Protocol 2.1.8. Sample as well as electrode buffer is Soln. D.

References

Reisfeld TA, Lewis UJ, Williams DE (1962) Nature 195:281

2.1.10 Affinity Electrophoresis

The principle of affinity electrophoresis is a nondenaturating ("native") electrophoresis using a matrix containing a ligand or receptor interacting with some of the separated macromolecules resulting in changed mobility of the complex (ligand-ligate interaction). It is possible to analyze whether a protein (e.g., glycoprotein) interacts generally with another macromolecule (e.g., lectin; cf. Table 2.19). By variation of the ligand or ligate concentration it is possible to estimate the dissociation constant of the complex.

Usually a non-covalent immobilized reactant is incorporated into the gel and the electrophoretic conditions are chosen in that way that a significant motion of this reactant does not occur.

Especially in the case of low molecular mass ligands these ligands are covalently attached to the gel matrix, e.g., by copolymerization of allylesters of mono- or oligosaccharides.

A special case of affinity electrophoresis is immunoelectrophoresis (cf. Chap. 4.10), which analyzes antibody-antigen interactions.

Horizontal Affinity Electrophoresis in Agarose Gels

Solutions/Reagents

A 1.5% agarose standard EEO (w/v) dissolved in ddH_2O for 2–3 h at about 80 °C
B barbital-acetate buffer I = 0.05, pH 8.6 (cf. Protocol 7.4.1)
PBS

The ligand is dissolved in Soln. B with a concentration of 0.2–1 mg/ml (for a 10 × 10-cm plate 5 ml are needed). Heat this solution to 60 °C and mix with an equal volume of agarose solution A, also heated to 60 °C. Pour the mixture onto a horizontally straightened glass plate. Introduce the comb and allow the gel to solidify at room temperature.

Mix the samples with Soln. B in a 1:1 ratio; supplement some bromophenol blue.

Put the plate on the cooling plate of a horizontal electrophoresis apparatus. Fill the buffer chambers with 1:1 diluted Soln. B (= electrode buffer) and connect the electrode buffer to the gel by filter paper bridges, wetted with electrode buffer. The filter paper covers the gel for about 5 mm. Remove carefully air bubbles between papers and gel.

Remove the comb and fill the samples, 5–10 μl each, into the pockets. It is recommended to cover the gel with a glass plate being on the filter bridge.

Cool the gel to about 10 °C, and then run the electrophoresis with cv of 8–15 V/cm. Proteins as well as bromophenol blue move to the anode (+).

Capillary blotting

In the case of protein ligands the detection of ligates and ligand-ligate complexes is done by blotting. After finishing the electrophoresis, remove the upper glass plate and the buffer bridges and cover the gel with a dry blotting membrane of same size (remove air bubbles).

The membrane is covered with a 3 to 4 mm stack of dry filter paper. Finish the sandwich with a glass plate and wrap all with plastic foil. After incubation overnight in a refrigerator remove the membrane, wash with PBS and block with an appropriate buffer, e.g., 0.5% BSA in PBS, 30 min.

Identify the ligate by specific antibodies as described in Protocol 2.5.4 (Western blot).

References

Heegard HH, Bjerrum OJ (1991) Anal Biochem 195:319

Takeo K (1987) In: Chrambach A, Dunn MJ, Radola BJ (eds.) Advances in electrophoresis, vol. 1, pp 229–279, VCH, Weinheim

Taketa K (1991) J Chromatogr 569:229

2.1.11 Two-Dimensional Polyacrylamide Gel Electrophoresis (2D-PAGE; IEF followed by SDS-PAGE)

Generally, two-dimensional gel electrophoresis (2D electrophoresis) is the sequential separation of a protein mixture in an electrical field, whereas the conditions of the first separation are different from those of the second one, running rectangular to the first. First and second dimension may differ in acrylamide concentration, pH, buffer composition, by addition of detergents or urea, or type of carrier (polyacrylamide or agarose). Mostly the combination of isoelectric focusing (IEF) for the first separation and SDS-PAGE for the second is described as "2D-PAGE." When using soluble carrier ampholytes in IEF for the first dimension, the method is designated as O'Farrell technique; it is named IPG-Dalt when ampholytes are co-polymerized (immobilized) within the polyacrylamide matrix (Görg 1988).

After finishing the 2D-PAGE, proteins are separated both by their isoelectric points (pI) and molar masses (M_r).

The main advantage of the IPG-Dalt protocol is a much higher reproducibility in gel preparation and running conditions, and mostly higher resolution of the sample.

2.1.11.1 First Dimension: Isoelectric Focusing (IEF)

Preparation of Immobiline Gels

Solutions/Reagents

A 29.1% acrylamide (w/v), 0.9% N,N′-methylene bisacrylamide (w/v) (30% T, 3% C), stored with 10% humid, neutral mixed-bed ion exchanger (w/v) in ddH_2O

B 10% (v/v) TEMED in ddH_2O

C 10% (w/v) ammonium persulfate in ddH_2O (prepare freshly)

D ***rehydration solution***[10]: 8 M urea, 1% CHAPS (w/v), 15 mM DTT, 0.2% ampholyte pH 3–10 (v/v)

Volumes for a IPG (IEF) gel 180 × 180 × 0.5 mm (glass plate 200 × 200 mm; total volume 16.2 ml) are given in Table 2.10.

Give some drops of water onto a clean 200 × 200 mm glass plate. Place a piece of GelBond foil, hydrophilic side upward, on the plate (it is recommended to draw a thin line with a pencil on one side of the foil for easier identification of, for example, the acidic side).

Table 2.10. Casting protocol for Immobiline gels with linear pH gradient

Solution (µl)	pH 4–9 %T = 4, %C = 3		pH 4–12 %T = 5, %C = 3	
	Acidic	Basic	Acidic	Basic
Immobiline pK 3.6	448	79	513	–
Immobiline pK 4.6	127	229	190	40
Immobiline pK 6.2	125	194	172	111
Immobiline pK 7.0	12	160	159	56
Immobiline pK 8.5	135	38	26	282
Immobiline pK 9.3	119	358	28	118
Immobiline pK 10.0	–	–	22	176
Immobiline pK > 13	–	–	–	287
Solution A	1080	1080	1350	1350
Glycerol (g)	2	–	2	–
H_2O	4396	5908	3910	5584
Mix, adjust pH 7.0 with 4 N NaOH and 4 N acetic acid, respectively				
Solution B	54	54	54	54
Solution C	32.4	32.4	32.4	32.4

Data according to GÖRG (2003)

[10] 24 g urea and 0.5 g mixed-bed ion exchange resin are filled up with H_2O to 50 ml. The mixture is stirred for 10 min at RT, than the resin is filtered off. CHAPS, DTT, and ampholyte are added in the indicated amount to the clear urea solution.

Expel the excess water with a photo roller. Place a second glass plate bearing the U-frame (for 0.5 mm distance) on top and clamp the cassette together. Cool this sandwich for 30 min in a refrigerator.

Give 8.1 ml of the acidic solution (mixture as given in Table 2.10 without Soln. C) into the "ds" chamber of the gradient mixer, open the connecting valve for a moment to let air out, then pour 8.1 ml of the basic solution (mixture as given in Table 2.10 without Soln. C) into the "ls" chamber. Add Soln. C to start polymerization, switch on the magnetic stirrer, and open both the valves.

When the cassette is filled after 2–3 min, allow the gradient to equilibrate for about 15 min at RT. The polymerization is completed for 1 h at 50 °C or overnight at RT.

Remove the IPG gel from the cassette and wash five to six times in a glass tray for 10 min each with 500 ml deionized water. Equilibrate the gel in 2% glycerol (w/v) for 30 min at RT, and then dry it overnight in a dust-free cabinet.

Cover the dry gel with a protective plastic foil, cut it into 5 mm strips and store airtight at −20 °C (strips are stable for several months).

Rehydration of Immobiline gels

Pour 300–400 µl of rehydration buffer D[11] into a small tray. Peel off the protective foil and insert the IPG gel face down into the tray. Cover gel and rehydration buffer with 1–2 ml silicone oil, close the lid of the tray, and allow swelling overnight at RT by gentle agitation.

Isoelectric focusing

Apply some drops of kerosene on the cooling block of the horizontal electrophoresis apparatus and apply the IPG strips with the acidic side to the anode. Make 1 cm broad strips from 2 mm thick filter paper. Wet the strips with deionized water (no ultrapure water!) and place them on top of the IPG strips as well as at the anodic (acidic) and the cathodic (basic) ends of the strip(s).

Use a silicone rubber frame (sample applicator) or a piece of filter paper (2 × 5 mm) for sample application about 5 mm from the anode or cathode and apply sample (concentration should not exceed 10 mg/ml), which is dissolved in rehydration buffer.

Place the electrodes and press them gently on the IEF electrode strips. Close the lid of the electrophoresis apparatus and switch on voltage. For 180 mm IPG strips the following conditions should be used (Table 2.11):

Temperature 20 °C, maximal current 0.05 mA/strip, maximal voltage 3500 V, maximal power 0.2 W/strip. For orientation the following schedule is used:

Remove the strips using tweezers after focusing. If the second run does not follow immediately, the strips may be frozen at −70 °C.

[11] Rehydration and sample application may be combined. Up to 5 mg/ml of sample can be added to the buffer D, if the M_r of the proteins are not to high or their pI to extreme.

Table 2.11. Running conditions for IPG gels

Sample in applicator	Sample applied during rehydration
1 h 150 V	1 h 150 V
2 h 300 V	2 h 300 V
1 h 600 V	
8 h 3500 V	8 h 3500 V

Marker proteins with known isoelectric point should be used to check the pH gradient. These proteins are stained by the common protocols (e.g., Coomassie Brilliant Blue, Protocol 2.3.1.2, or silver staining, Protocol 2.3.2).

If the IPG gels are not used for a second dimension run, the gels are further processed (stained, blotted) as described for SDS gels.

2.1.11.2 Second Dimension: SDS-PAGE (Acrylamide Gradient Gel)

The discontinuous Laemmli SDS-PAGE with polyacrylamide concentration gradient in the separating gel is mostly used for second dimension in 2D-PAGE (cf. Protocol 2.1.1). Of course, other electrophoretic systems are also applicable.

For convenience, prepare the gel in vertical position but run the second dimension on a horizontal apparatus.

Form a cassette by glass plates 1 – 2 cm wider than the length of the IPG strip and with 0.75 to 1 mm thick spacers. Cover one plate with a thoroughly water-washed GelBond PAGE sheet (hydrophilic surface to the lumen of the cassette).

Prepare the gel as described in Protocol 2.1.1, i.e., first a separation gel followed by a stacking gel without sample application slots.

Solutions/Reagents

- E 6 M urea, 30% glycerol (w/v), 2% SDS, 50 mM Tris-HCl, pH 8.8, 0.125 mg/ml bromophenol blue in ddH_2O
- E′ 0.2 M iodoacetamide in buffer E

First equilibrate the IPG strip for 15 min in sufficient buffer E, then for an additional 15 min in buffer E′. Rinse the strip with ddH_2O after equilibration and suck off carefully remaining water by filter paper.

Place the IPG strip face down onto the stacking gel, about 5 mM apart of the cathode paper bridge. Place silicone rubber sample applicators for molar mass marker proteins at one or both sides of the IPG strip.

Run the electrophoresis as described in Protocol 2.1.1. Cool the gel to about 15 °C during the run. When the tracking dye bromophenol blue has reached the interphase between stacking and separation gel, stop electrophoresis. Remove the IPG strip and move the electrode paper strip to that place where the IPG strip was. Continue electrophoresis with 10 – 15 V/cm until bromophenol blue arrives at the anodic paper bridge.

Identify the protein spots as usual, i.e., by staining, autoradiography, gel overlay, or Western blot. In the latter case the gel must be separated from the GelBond foil prior to electrotransfer. For this purpose a film remover (Görg 2003, Fig. 19) is used: The gel is placed on the cylindrical remover with foil down, clamped on an edge, and a thin stainless steel or nylon wire is pulled between foil and gel towards to your body. Cover the gel with the wetted blotting membrane (cf. Protocol 2.4.3) and transfer membrane as well as gel to the blotting apparatus.

Western blot of 2D gel

References

O'Farrel PH (1975) J Biol Chem 250:4007
Dunn MJ, Burghes AHM (1983) Electrophoresis 4:97; 4:173
Görg A, Postel W, Günther S (1988) Electrophoresis 9:531
Righetti PG (1990) Immobilized pH gradients: theory and methodology. Laboratory techniques in biochemistry and molecular biology, vol. 20, Elsevier, Amsterdam
Westermeier R (1997) Electrophoresis in practice: a guide to methods and applications, 2nd ed. VCH, Weinheim
Görg A, Obermaier C, Boguth G, Harder A, Scheibe B, Wildgruber R, Weiss W (2000) Electrophoresis 21:1037
Görg A (2003) Two-dimensional electrophoresis with immobilized pH gradients for proteome analysis. http://www.weihenstephan.de/blm/deg

2.2 Agarose and Paper Electrophoresis

2.2.1 Non-denaturating Nucleic Acid Electrophoresis

This system separates double-stranded DNA (dsDNA) fragments with a length of 70 × 80 000 base pairs (bp) in gels of 3% to 0.1% agarose. For analysis of smaller-fragments (6000 to 1000 bp) PAGE systems are described in literature with 20 – 3% polyacrylamid.

During sample preparation and electrophoresis DNA will not be denaturated, i.e., the double strand and higher elements of structure stay unaffected. To avoid unwanted further fragmentation all manipulations have to be done without possible contaminations by fingerprints, droplets of saliva or microorganisms. Sterilized materials are recommended.

Since protocols for electrophoresis used in DNA sequence analysis are given, for example, by Sambrock et al., these specialized methods are not outlined in this chapter.

Solutions/Reagents

A 200 mM Tris-acetate, 10 mM EDTA, pH 8.2 (TAE buffer[12])
B 0.1 – 3.0% agarose (low endoosmosis) (w/v) in 1/5 buffer A

[12] Fiftyfold stock of TAE buffer see Chap. 7.4

C 50% glycerol (v/v), 0.3% bromophenol blue (w/v) in buffer A
D 1 mg/ml ethidium bromide[13] in ddH_2O

***Caution!** Wear gloves! Ethidium bromide is carcinogenic.*

The agarose (Soln. B) is molten in a boiling water bath and cooled to about 50 °C. Soln. D is added to a final concentration of 0.5 µg/ml[14]. The liquid agarose is poured into the gel chamber of a horizontal electrophoresis apparatus and the comb is inserted.

The sample is dissolved in a buffer with low ionic strength and mixed with 1/5 of its volume of Soln. C. The amount of DNA should be about 50 µg per square centimeter of slot area.

Electrode buffer is 1:5 diluted buffer A, adjusted to pH 8.2 and supplemented with D to a final concentration of 0.5 µg/ml.

Electrophoresis is performed with 20–60 V cv overnight and the applied voltage is inverse proportional to the mean DNA size.

The addition of ethidium bromide (EtBr) or propidium iodide (PI) allows the identification of DNA bands immediately after run on a transilluminator. Intercalation of the dyes into the double-stranded DNA yields fluorescence (EtBr: excitation wavelength λ_{ex} 302 or 366 nm, emission wavelength λ_{em} 590 nm; PI: λ_{ex} 530 nm, λ_{em} 620 nm).

Separated DNA may be eluted from cut pieces of gel or transferred to membranes by SOUTHERN blot.

References

Rickwood D, Hames BD (eds.) (1984) Gel electrophoresis of nucleic acids: a practical approach. IRL Press, Oxford

Sambrook J, Fritsch EF, Maniatis T (1989) Molecular cloning: a laboratory manual, 2nd ed., vol. 1, chap. 6, Cold Spring Harbor Laboratories Press

Jany K-D, Hahn H (1991) In: Bertram S, Gassen G (eds.) Gentechnische Methoden. pp. 39–43. Fischer, Stuttgart

2.2.2 Denaturating Nucleic Acid Electrophoresis

General differences between DNA and RNA in denaturating electrophoresis do not exist; therefore, the following protocols may be used, as well as for DNA and RNA.

Solutions/Reagents

A 10 mM sodium phosphate, pH 6.5–6.8
A′ 10 mM sodium phosphate, pH 7.4, 1.1 M formaldehyde

[13] Instead of EtBr use less harmfull fluorescent dyes, e.g., propidium iodide (PI) or SYBR Green (cf. Haugland RP (1996) Handbook of fluorescent probes and research chemicals, 6th ed., Chap. 8, Molecular Probes, Eugene, Oregon; https://catalog.invitrogen.com)

[14] If ethidium bromide is omitted as well as in gel and electrode buffer, the DNA sample is supplemented with 5 µl of Soln. D per milliliter and incubated for 10 min at RT.

A″ 0.4 M Tris · HCl, 20 mM sodium acetate, 1 mM EDTA, pH 7.4
B 6% glyoxal (w/v) in Soln. A
C 0.8–1.5% agarose (w/v) in Soln. A[15]
C′ 0.8–1.5% agarose (w/v) in Soln. A′ [14]
C″ 0.8–1.5% agarose (w/v) in 6 M urea, 15 mM iodoacetate, pH 7.4 [14]
D 50% glycerol (v/v), 0.3% bromophenol blue in ddH_2O
D′ 50% formamide (v/v), 6.5% formaldehyde (w/v), 0.5 mM EDTA, 10-mM sodium phosphate, pH 7.4
E′ 25% glycerol (v/v), 0.5% SDS (w/v), 0.025% bromophenol blue (w/v), 25 mM EDTA in ddH_2O

Variant 1: Glyoxal Denaturation

A gel of 2–8 mm thickness is made from Soln. C in a horizontal slab apparatus.

The RNA sample is incubated in Soln. B for 1 h at 50 °C. After cooling, the solution is mixed with 1/10 of its volume with Soln. D and loaded into the slots of the gel.

Electrophoresis is run for about 18 h at RT with constant voltage of 25–30 V. Electrode buffer is Soln. A, which circulates between the electrode chambers during electrophoresis.

After the end of the electrophoresis, the nucleic acids are stained with aqueous EtBr (0.5 µg/ml) and visualized on a UV transilluminator.

Variant 2: Formaldehyde Denaturation

A gel of 2 to 8 mm thickness is made from Soln. C′ in a horizontal slab apparatus.

RNA samples dissolved in Soln. D′ are heated to 65 °C for 5 min, cooled and mixed with 1/5 of their volumes of Soln. E′.

Electrophoresis is done in a hood with 35–60 V (cv) using circulating Soln. A′ as electrode buffer.

Staining and identification of bands are described in Variant 1.

Variant 3: Urea gel

A gel of 2–8 mm thickness is made from Soln. C″ in a horizontal slab apparatus. Because urea decomposes slightly during heating, it should be avoided to melt it several times or to keep it melted for longer periods.

The RNA samples are dissolved in any buffer, the ionic strength of which has to be between 0.01 and 0.2 and which may content detergents.

Run electrophoresis with 5 V/cm (cv) for 3–6 h using the electrode buffer A″.

Staining and identification of bands are described in Variant 1.

[15] The agarose concentration depends on the average size of nucleic acid fragments: the smaller the fragments, the higher the concentration.

Single-stranded nucleic acids are easily and with high efficiency transferred to membranes for hybridization experiments.

References

Locker J (1980) Anal Biochem 98:358
Rickwood D, Hames BD (eds.) (1984) Gel electrophoresis of nucleic acids: a practical approach. IRL Press, Oxford
Voytas D (2000) Current protocols in molecular biology. 2.5A. Wiley, New York

2.2.3 Identification of Phosphoamino Acids (Paper Electrophoresis)

Amino acid paper electrophoresis has sufficient sensitivity and reproducibility for amino acid analysis. If analysis by HPLC is not available in or around your lab, paper electrophoresis may be an alternative.

Solutions/Reagents

A glacial acetic acid/pyridine/ddH_2O 50:5:945 (v/v/v), pH 3.4
B ninhydrin spray: 0.3 g ninhydrin dissolved in 100 ml n-butanol and supplemented with 3-ml glacial acetic acid
6 N hydrochloric acid

Electrophoresis is done on ready-to-use cellulose plates or on chromatography paper (e.g., Whatman No. 3MM, Schleicher and Schuell No. 2040b, Macherey and Nagel MN) of 20 cm length.

The protein or peptide sample is hydrolyzed in a closed tube at 110 °C for 2 h[16]. After hydrolysis the samples are lyophilized, dissolved in a small volume of ddH_2O and lyophilized again.

The start area on the plate (strip) is marked with a pencil at the cathodic side. If several samples should be analyzed in the same run, the start areas should have a distance of 1.5–2 cm. A plastic sheet covers the starting zone and the plate is sprayed with Soln. A until it is humid but not wet.

The absolutely dry, lyophilized samples are dissolved in ddH_2O to a concentration of 0.2–1 mg/ml (1.5–5 pmoles/µl) per amino acid. Up to 5 µl of the samples are spotted on the marked start.

The connection between electrode chambers of a horizontal electrophoresis apparatus filled with buffer A and the cellulose support is made by wetted paper bridges. To avoid liquid moving, the buffer level must be the same in both chambers. A glass plate, laying on the paper bridges, covers the separation plate cooled to 0–4 °C.

Important! *Take care for effective and constant cooling during the electrophoresis.*

[16] The hydrolysis conditions of phosphoamino acids with respect of stability and reproducibility of quantitative determination are discussed by Bylund and Huang (1976). Anal Biochem 73:477).

The electrophoresis is run at 1000 V cv for 30–45 min. After finishing, the cellulose sheet is dried in a stream of warm air. For visual detection of amino acids the cellulose sheet is sprayed with Soln. B and incubated in a drying oven at 110–120 °C until blue spots appear. Autoradiography or a PhosphoImager detects ^{32}P-labled amino acids.

Phosphorylated amino acids move in the order serine phosphate > threonine phosphate > tyrosine phosphate.

For quantitative determination the corresponding spots and a blank area of same size are cut out. The cellulose is burned with concentrated nitric acid and the residue is used for phosphate determination according to Protocol 1.3.2, enzymatically by coupled optical test and in the case of ^{32}P-labeling by counting radioactivity.

References

Hunter T, Sefton BM (1980) Proc Natl Acad Sci USA 77:1311

2.3 Aid in Electrophoresis

A lot of companies offer numerous kits for calibration and staining of electropherograms. These kits simplify working with gels, but though they are optimized with respect to their composition, they are mostly expensive and self-made reagents give the same, and in some cases better, results. Besides, the following protocols gives a better understanding of mechanisms of calibration and staining.

2.3.1 Marker Dyes for Monitoring Electrophoresis

2.3.1.1 Anodic Systems

Solutions/Reagents

0.02 g bromophenol blue in 0.25 ml ethanol and 0.02 g Pyronin Y (Pyronin G) in 0.25 ml ddH_2O separately dissolved, combined and supplemented with 0.5 ml glycerol or 1 ml 50% sucrose (w/v) in ddH_2O, filled up with ddH_2O to 2 ml

0.05 µl of this dye solution are added per 1 µl sample solution of all SDS-PAGE systems with a pH of separation gel > 5. The color of bromophenol blue solution switches from yellow-brown to blue-violet. Better than bromophenol blue, especially at separation of small polypeptides, is Orange G (C.I. 16230), because it moves directly with the electrophoresis front. Unfortunately, Orange G is not as easily visible as bromophenol blue or Pyronin Y.

2.3.1.2 Cathodic Systems

Solutions/Reagents

0.5% basic fuchsin (w/v), 50% sucrose (w/v) in ddH_2O

0.05 µl of this solution are added per microliter sample. Further addition of sucrose or glycerol is not necessary.

Other dyes for cathodic (acidic) electrophoresis systems are Pyronin Y, 1-napthyl red and methyl red, the latter only for systems with pH below 3.

2.3.2 Marker Proteins for the Polyacrylamide Gel Electrophoresis

Prerequisites for the use of a protein or polypeptide as calibrator in PAGE are the accessibility in a pure, definite state, its known

Table 2.12. Marker proteins for polyacrylamide gel electrophoresis

Protein	M_r in kD	
	With SDS	Without SDS
Glucagon		3.483
Insulin (pig)		5.733
Insulin A chain		2.533
Insulin B chain		3.495
Trypsin inhibitor (Aprotinin; bovine)		6.5
Cytochrome c (pig)		12.5
Lysozyme (hen egg)		14.3
α-Lactalbumin (bovine)		14.4
Hemoglobin (bovine)		16
Myoglobin (horse)		16.95
Myoglobin. BrCN-fragment III		2.512
Myoglobin. BrCN-fragment II		6.214
Myoglobin. BrCN-Fragment I		8.159
β-Lactoglobulin (bovine)		18.4
Trypsin inhibitor (soy bean)		20.1
Trypsinogen (bovine)		24
Chymotrypsinogen A (bovine)		25.7
Carbonic acid anhydrase (bovine)		29
Pepsin (pig)		34.7
Aldolase (rabbit)		40
Ovalbumin (hen egg)		42.7
Glutamate dehydrogenase (bovine)		53
Pyruvate kinase (rabbit)		57.2
Serumalbumin (bovine)		66.25
Transferrin (human)		76
Ovotransferrin (hen egg)		76–78
Phosphorylase b (rabbit)		94
β-Galactosidase (*E.coli*)		116.3
Lactate dehydrogenase	36	140
α-Macroglobulin (bovine)		170
Myosin (rabbit)		205–215
Catalase	57.5	232
Thyroglobulin		305
Ferritin (horse)	18.5 + 220	440

Table 2.13. Calibration kits for polyacrylamide gel electrophoresis (M_r in kD)

Peptide marker kit	M_r
Myoglobin peptides	16.949
	14.404
	10.7
	8.159
	6.214
	2.512

LMW kit	M_r	mg/ml
Phosphorylase b	96	0.67
Serum albumin	67	0.83
Ovalbumin	43	1.47
Carbonic acid anhydrase	30	0.83
Trypsin inhibitor	20	0.80
α-Lactoglobulin	14	1.21

HMW Native Kit	M_r	mg/ml
Thyroglobulin	669	0.76
Ferritin	440	0.5
Katalase	232	0.36
Lactate dehydrogenase	140	0.48
Serum albumin	66	0.4

HMW SDS Kit	M_r	mg/ml
Myosin	210	0.25
α_2-Macroglobulin	170	1.00
β-Galactosidase	116	0.16
Transferrin	90	0.17
Glutamate dehydrogenase	53	0.18

Manufacturer: GE Healthcare; formerly Amersham Biosciences

MW SDS-70	M_r	mg/ml
Serum albumin	67	1.5
Ovalbumin	43	1.5
Pepsin	35	5.0
Trypsinogen	24	2.0
α-Lactoglobulin	17	1.0
Lysozym	14	1.0

MW SDS-200	M_r	mg/ml
Carbonic acid anhydrase	30	1.0
Ovalbumin	43	1.5
Serum albumin	67	1.5
Phosphorylase b	96	1.0
β-Galactosidase	116	1.0
Myosin	204	1.7

Manufacturer: Sigma Aldrich

SDS-PAGE Rainbow marker	Low M_r	High M_r	Color
Myosin		220	Blue
Phosphorylase b		97.4	Brown
Serum albumin		66	Red
Ovalbumin	46	46	Yellow
Carbonic acid anhydrase	30	30	Orange
Trypsin inhibitor (soy bean)	22	21.5	Green
Lysozym	14.3	14.3	Purple
Aprotinin	6.5		Dark blue
Insulin B chain	3.4		Blue
Insulin A chain	2.4		Blue

Manufacturer: BioRad

molar mass, and a normal movement in electrophoresis, e.g., linear correlation with respect to other standard proteins. Not all proteins given in Table 2.12 fulfill these criteria; therefore, an additional table (Table 2.13) is given with proteins often used in SDS-PAGE.

Since the obtained optical density (intensity of stain) per mass unit of a protein differs depending on its nature and the staining method, for some commercial available protein mixtures the composition is given, which produces staining a homogeneous picture by Coomassie.

About 1–5 µl of the pre-mixed ready-to-use protein standards per lane are recommended in SDS-PAGE combined with Coomassie staining.

2.3.3 Covalently Colored Marker Proteins

Protein molar mass standards covalently labeled with dyes are valuable when the electrophoresis is followed by a Western blot or when electrophoresis has to be monitored. Furthermore, procedures of gel staining alter the geometry, and assignment of bands on blot to marker bands within the gel is sometimes difficult.

The electrophoretic mobility of the labeled proteins obtained by this protocol is practically not different from that of the unlabeled proteins.

Solutions/Reagents

A 0.1 M borate buffer, 5% SDS (w/v), pH 9.0

B 10 mM dimethylamino-azobenzen-4′-sulfonyl chloride (dabsyl chloride, M_r 323.8) in acetone

C ***electrophoresis sample buffer***: 50 mM Tris-HCl, 4% SDS (w/v), 5% 2-mercaptoethanol (v/v), 10% sucrose or glycerol (w/v), pH 6.8.

The respective marker proteins, single or mixed, are dissolved in Soln. A to a concentration of 20 mg/ml at RT. Then 0.5–1 vol. of Soln. B are added and the mixture is heated to 60 °C for 5 min. After cooling, the labeled proteins are lyophilized without further purification and dissolved again in Soln. C to a concentration of 20 mg/ml. If the assay was larger than 0.1 ml, the product is aliquoted and stored at −20 °C.

The remaining dabsyl chloride is not removed, since it reacts with Tris of Soln. C to a dye acting as tracking dye like bromophenol blue.

1 to 2 µg per lane of the obtained labeled marker proteins each are applied.

Several manufacturers distribute ready-to-use colored marker proteins (cf. example in Table 2.13).

References

Tzeng M-C (1983) Anal Biochem 128:412
Lin J-K, Chang J-Y (1975) Anal Chem 47:1634

2.4 Staining Protocols

2.4.1 Staining with Organic Dyes

Because the ability to dye one protein may differ from that of another in an electropherogram, an unknown protein mix should become stained by different methods. Some of the procedures can be done sequential in the same gel, e.g., additional Coomassie staining after silver staining.

A synopsis of dyes used for visualization after electrophoresis is given in Table 2.14.

Densitometry

Staining can also be used for quantifying proteins in gels. The gel is scanned and the obtained picture is calculated by densitometric software. The amount of protein is proportional to the peak area of the whole band. Since the reaction of a protein with a dye can vary in a broad range (row "Bradford" in Table 1.1 corresponds to Coomassie staining), calibration of the intensity is only possible when the same or a very similar protein is used.

Qualitative interpretation of stained gels is often misleading, since a blank area has not to be protein-free and an intensive band has not to reflect a large amount.

Table 2.14. Selected dyes suitable for electrophoresis

Dye	Staining of		Color index
	Proteins	Nucleic acids	
Acridine Orange		x (RNA)	46005
Amido Black 10B	x		20470
8-Anilino-1-naphthalenesulfonic acid (ANS)	x		
Bismarck Brown R	x (together with Coomassie)		21010
Coomassie Blue G250	x		42655
Coomassie Blue R250 and R350	x (more sensitive than G250)		42660
Coomassie Violett	x (for IEF)		42650
Ethidium bromide		x (fluorescence)	
Fast Green FCF	x		42053
Methylen Blue		x (RNA)	52015
Methyl Green		x (native DNA)	42590
Ponceau S	x (Western blots on NC)		27195
Stains All	x	x	
Sudan Black	x (for lipoproteins)		26150
Sulforhodanin B	x (Western blots on PVDF)		45100

2.4.1.1 Amido Black 10 B

(Naphthol Blue Black 6 B)

Solutions/Reagents

A Dissolve 0.5 g Amido Black in 30 ml methanol. 10 ml of glacial acetic acid are added and the solution is filled up to 100 ml with ddH_2O.

B *destaining solution*: 10% acetic acid (v/v), 30% methanol or nondenaturated ethanol or isopropanol (v/v) in deionized water.

Gels may be fixed in Soln. B, i.e., a previous fixation by TCA or 5-sulfosalicylic acid is mostly not necessary.

Duration of staining is about 30–45 min at RT for a 1 mm thick gel.

After staining, the Amido Black solution is poured off (Soln. A can be used several times) and the gel is destained by several changes of Soln. B until the background becomes clear. If destaining was too strong, i.e., weak bands disappeared, staining and destaining can be repeated. Regeneration of Soln. B is possible by filtration over charcoal of mixed-bed ion exchanger.

Agitate the gel during staining and destaining. Elevated temperature (40–50 °C) reduces staining and destaining times considerably.

Densitometry of stained gel should be done at 620 nm.

2.4.1.2 Coomassie Brilliant Blue R250 and G250

(Brilliant Blue R and G; Acid Blue 83 and 90; C.I 42655 and C.I. 42660)

Solutions/Reagents

A 0.05 g of dye are dissolved in 30 ml methanol or non-denaturated ethanol or isopropanol. 10 ml of glacial acetic acid are added and the solution is filled up to 100 ml with ddH_2O. If the color does not change remarkably, the solution can be reused.

B *destaining solution*: 10% acetic acid (v/v), 30% methanol or non-denaturated ethanol or isopropanol (v/v) in deionized water

B′ *destaining solution*: 10% acetic acid (v/v), 50% methanol or non-denaturated ethanol or isopropanol (v/v) in deionized water

Both dyes, especially R250, are used mostly in staining gels and should be preferred. A second staining with Coomassie is recommended after silver staining.

Fixation, staining, and destaining as described in Protocol 2.4.1.1.

If Coomassie R250 (R – reddish, more sensitive than G250)[17] is used, a TCA fixation should be avoided. If the gel was fixed with

[17] Three R-dyes are described: R100; R250; and R350. They have the same C.I. number 42660, but the sensitivities are in the ratio of 1:2.5:3.5 (information by Amersham Biosciences).

TCA, equilibrate the gel twice for 20 min in Soln. B prior to staining. Destaining has to be done alternating with Soln. B and B′.

The destained gel may be stored in Soln. B for several days, but weak bands fade slowly.

Densitometry of stained gel should be done at 590 nm.

Variant for thin gels

By modification of Soln. A (0.05 g of dye are dissolved in 10 ml glacial acetic acid and filled up with deionized water to 100 ml), staining is done at 60 °C and destaining is done with 10% acetic acid in deionized water.

2.4.1.3 Coomassie Brilliant Blue R250 Combined with Bismarck Brown R

Gels are stained very sensitive and quick by combination of Coomassie Brilliant Blue R250 with Bismarck Brown R (Basic Brown 4; C.I. 21010). In our opinion, this is the best staining method using organic dyes.

Solutions/Reagents

A 0.2 g Coomassie Brilliant Blue R250 are dissolved in 40 ml methanol or non-denaturated ethanol; then 10 ml glacial acetic acid are added and the solution is filled up to 100 ml with deionized water. Stable at RT

B 0.05 g Bismarck Brown R are dissolved in 40 ml methanol or non-denaturated ethanol; then 10 ml glacial acetic acid are added and the solution is filled up to 100 ml with deionized water. Stable at RT.

C 1 vol. A is mixed with 0.75 vol. B immediately prior use.

D ***destaining solution***: 10% acetic acid (v/v), 30% methanol or non-denaturated ethanol or isopropanol (v/v) in deionized water

The gel is fixed for 5–10 min in Soln. D after electrophoresis and stained with Soln. C (about 20 ml for a 8 × 6 × 0.1 cm gel) for 20 min with agitation. Flocculation of dye does not influence staining.

Destaining is done with Soln. D.

Sensitivity is about twice the standard protocol without Bismarck Brown.

References

Choi J-K, Yoon S-H, Hong H-Y, Choi D-K, Yoo G-S (1996) Anal Biochem 236:82

2.4.1.4 Fast Green FCF

(Food Green 3; C.I. 42053)

Solutions/Reagents

A 0.25 g Fast Green are dissolved in 30 ml ethanol and 10 ml glacial acetic and filled up to 100 ml with deionized water

B ***destaining solution***: 10% acetic acid (v/v), 30% methanol or denaturated ethanol (v/v) in deionized water

Stain the gel for 1 – 3 h. Further comments see Protocol 2.3.1.1.

Staining with Fast Green is partial suitable for quantitative densitometry at 625 nm.

2.4.1.5 Stains All

(4,4′,5,5′-Dibenzo-3,3′-diethyl-9-methylthia carbocyanine bromide)

Solutions/Reagents

A *stock solution*: 0.01 g dye is dissolved in 10 ml pure formamide. It is stored airtight in the dark.

A′ *working solution*: Immediately before using, Soln. A is diluted 1:20 with Soln. B. The working dilution can be used only once.

B 50% formamide (v/v) in deionized water.

Acidic fixed gels are equilibrated three times for 20 min in Soln. B. Then the gel is stained with Soln. A′ for 1 h in the dark followed by destaining in Soln. B.

Stains All is light-sensitive and bleaches slowly at the air. The stained gel should dry in a vacuum gel dryer immediately after destaining and documentation.

The main advantage of Stains All are different colors produced by proteins and nucleic acids.

2.4.1.6 Staining of Proteolipids, Lipids, and Lipoproteins

Solutions/Reagents

A 0.5 g Sudan Black B (Solvent Black 3, Fat Black HB, C.I. 26150) is dissolved in 20 ml acetone, and then 15 ml of glacial acetic acid are added followed by 80 ml ddH_2O. The solution is centrifuged for 15 min at RT with 3000 × g, the supernatant is used for staining.

B 15% acetic acid (v/v), 20% acetone (v/v) in ddH_2O.

After electrophoresis the gel is fixed either in the common acetic acid/methanol/H_2O mixture or in Soln. B, but never in aqueous TCA.

Staining is done by incubation with a tenfold gel volume of Soln. A overnight. Use several changes of Soln. B for destaining. Remaining colloidal dye on the surface of the gel is wiped off with a wetted tissue.

References

Maurer HR (1971) Disc electrophoresis, p. 76, W. de Gruyter, Berlin

2.4.2 Silver Staining of Proteins in Gels

The following procedures are well improved and easy to perform. Nevertheless, in some cases variations of the protocols concerning silver binding and/or reduction can give better or other results, because proteins bind silver differently and are observed with varying colors. Some hints are given in the cited papers.

2.4.2.1 Citrate/Formaldehyde Development

Solutions/Reagents

A 10% acetic acid (v/v), 30% methanol (v/v) in ddH_2O
B 15% methanol (v/v) in ddH_2O
C 5% glutaraldehyde (v/v) in ddH_2O (usable for several times)
D 0.7 ml of 25% ammonia and 2.0 ml 1 N NaOH are filled up to 16.0 ml with ddH_2O, then 2.0 ml 10% $AgNO_3$ (w/v) are added dropwise with shaking. The precipitate formed by a drop has to dissolve completely before giving the next. If the last drops are not dissolved completely, some ammonia can be carefully added. The resulting solution is pale brown. When the addition of silver is complete, ddH_2O is added to 100 ml.
The solution is prepared freshly before using. After use, the solution has to be destroyed by hydrochloric acid.
E ***reduction solution***: 0.025 g citric acid and 0.185 ml 27% formaldehyde in 100 ml ddH_2O (stabilize for a longer period if frozen)
F 0.5% acetic acid (v/v) in H_2O

The gel to be stained should not be thicker than 1 mM for that diffusion of the reactants is quickly. The volumes of used solutions should be at least the tenfold of the gel volume.

The gel is moved gently during all steps. Especially immediately before, during, and after the application of Soln. D the gel must not be touched with unprotected fingers (use gloves and/or tweezers) or come in contact with chloride-containing solutions.

After electrophoresis the gel is fixed in Soln. A at least 20 min. Then it is incubated three times in Soln. B for 20 min each, 30 min in Soln. C, and three times in Soln. B again. Incubation in Soln. D succeeds for 30 min. Excess of glutaraldehyde is removed by three washings with ddH_2O for 5 min each.

Development is done with Soln. F. Pouring off Soln. F and add Soln. G when first bands appear to stop the reaction.

Overstained gels must be freed from acidic and/or silver residues by washing with several changes of ddH_2O. Then the staining can be reduced or removed according to Protocol 2.4.2.7. A new silver stain is possible, but you never get the same picture as at the first time; therefore, the weakening should be done only if absolutely necessary.

A second staining with Coomassie Brilliant Blue following the silver stain perhaps produces bands of proteins not or weak stainable by silver, because silver staining is, on the one hand, more sensitive than other procedures, but on the other hand, also not universal.

Second staining

For a second staining with Coomassie, the gel is equilibrated three times in Soln. F and then stained as described in Protocol 2.3.1.2.

Sometimes the electrophoresis buffer produces a high background. A significant, rapid increase of electric field strength, as well as reaction products of β-mercaptoethanol or DTT, produces pseudobands. The artifacts made by reductants can be excluded by using TCEP · HCl instead of sulfhydryl reductants (cf. Protocol 2.1.1).

2.4.2.2 Alkaline Development

A further variant of silver staining of proteins resulting in some other color of protein bands has been described by HEUKESHOVEN and DERNICK.

Solutions/Reagents

- A 10% (v/v) acetic acid, 30% (v/v) ethanol in H_2O
- B 6.8 g sodium acetate, 0.2 g $Na_2S_2O_3 \cdot 5H_2O$ (fixing salt) dissolved in 60 ml ddH_2O, then 0.5 ml 25% glutaraldehyde is added. 30 ml of ethanol are followed and ddH_2O is filled up to 100 ml.
- C 0.25% $AgNO_3$ (w/v) in ddH_2O, supplemented with 4 µl 37% formaldehyde per 10 ml immediately before use.
- D ***reduction solution***: 2.5% Na_2CO_3 (w/v) in ddH_2O, pH 11 to 11.5, supplemented with 2 µl 37% formaldehyde per 10 ml immediately before use.
- E 1% glycine (w/v) in ddH_2O

After electrophoresis, the gel is fixed in Soln. A for 30 min, then it is equilibrated at least for 30 min or overnight in Soln. B. Next the gel is washed with ddH_2O three times for 5 min and agitated in Soln. C for 20 min.

Give the gel into Soln. D. After 1 min, the reducing solution is rejected and fresh Soln. D is added to the gel. During this step the color of bands develops. When the bands are clear visible, Soln. E replaces Soln. D.

Washing three times with water for 5 min each and drying finishes the staining procedure.

References

Heukeshoven J, Dernick R (1985) Electrophoresis 6:103

2.4.2.3 Silver Staining Using Tungstosilicic Acid

Solutions/Reagents

- A 50% methanol (v/v), 10% acetic acid (v/v), 5% glycerol (v/v), 35% H_2O (v/v)
- B 2.5 ml 2% $AgNO_3$ (w/v) in ddH_2O and 2% NH_4NO_3 (w/v) in ddH_2O are mixed immediately before use; then 2.5 ml 10% tungstosilicic acid (w/v) in ddH_2O, 0.25 ml 27% formaldehyde, and 25 ml 5% anhydrous Na_2CO_3 (w/v) in ddH_2O are added and the mixture is filled up to 50 ml.
- C 5% acetic acid (v/v)

The gel is fixed with about a fivefold of its volume with Soln. A for 30 min. Then it is washed four times with water for 5 min each.

Put the gel into a fivefold volume of Soln. B and observe development of bands. Discard Soln. B after 10–15 min and stop formation of bands by Soln. C.

Store the gel in Soln. A or C, or dry it.

References

Information from BioRad

2.4.2.4 Silver Staining of Proteins: Formaldehyde Fixation

Colors and intensity of bands may differ after formaldehyde fixation compared with those obtained by glutaraldehyde fixation. Formaldehyde yields blackish and glutaraldehyde gives brownish bands.

Solutions/Reagents

A 10% acetic acid (v/v), 40% nondenaturated ethanol (v/v), 0.8% formaldehyde[18] (w/v) in ddH_2O

B 1.4 ml of 25% ammonia and 4.0 ml 1 N NaOH are filled up to 16.0 ml with ddH_2O, then 4.0 ml 10% $AgNO_3$ (w/v) are added dropwise with shaking. The precipitate formed by a drop has to dissolve completely before giving the next. If the last drops are not dissolved completely, some ammonia can be added carefully. The resulting solution is pale brown. When the addition of silver is complete, ddH_2O is added to 100 ml.
The solution is prepared freshly before using. After use, the solution has to be destroyed by hydrochloric acid.

C 0.05% citric acid (w/v), 0.185 ml of 27% formaldehyde [18] in ddH_2O

D 0.5% acetic acid (v/v) in ddH_2O

The gel is fixed for 1 h in Soln. A. Wash the gel at least four times with water; about 30 min for each wash and preferably overnight for the last wash. Agitate slowly with each wash. Stain the gel with at least five gel volumes Soln B (to cover the gel) for 30 min with shaking. Pour off the silver solution, wash with ddH_2O four times for 5 min each and develop with Soln. C as described in Protocol 2.4.2.1. Development is stopped after decanting Soln. C by Soln. D. The gel is stored in Soln. D until drying.

References

Ochs DV, McConkey EH, Sammons DW (1981) Electrophoresis 2:304
Yüksel KÜ, Gracey RW (1985) Electrophoresis 6:361

[18] Instead of aqueous formaldehyde (formalin) the required amount of paraformaldehyde is weighed out, dissolved in water followed by addition of the other components.

2.4.2.5 Silver Staining of Glycoproteins and Polysaccharides

This procedure is not solely specific for carbohydrate side chains of proteins. Unglycosylated proteins may also be stained. To identify glycosylated proteins, the sample should be run in at least two identical lanes: cut the gel and stain a lane with the common protein silver stain (Protocols 2.4.2.1 to 2.4.2.4) and the other lane by the described method. Compare pattern and intensity to identify glycoproteins. Glycosylated macromolecules are also stainable with SCHIFF's reagent (Protocol 2.4.4.1), but with less sensitivity.

Solutions/Reagents

- A 5% acetic acid (v/v), 40% nondenaturated ethanol (v/v) in ddH_2O
- B 0.7% sodium metaperiodate (w/v) in Soln. A
- C 1.4 ml of 25% ammonia and 4.0 ml 1 N NaOH are filled up to 16.0 ml with ddH_2O then 4.0 ml 10% $AgNO_3$ (w/v) are added dropwise with shaking. The precipitate formed by a drop has to dissolve completely before giving the next. If the last drops are not dissolved completely, some ammonia can be added carefully. The resulting solution is pale brown. When the addition of silver is complete, ddH_2O is filled up to 100 ml.
  The solution is prepared freshly before using. After use, the solution has to be destroyed by hydrochloric acid.
- D 0.025 g citric acid, 0.185 ml of 27% formaldehyde in 100 ml ddH_2O
- E 0.5% acetic acid (v/v) in ddH_2O

All steps have to be done in glass vessels cleaned with concentrated nitric acid and rinsed thoroughly with de-ionized water.

The gel is fixed after electrophoresis for at least 30 min, better overnight, in Soln. A. The fixing solution is poured off and the gel is agitated in Soln. B for 5 min. Wash the gel three times with ddH_2O for 15 min each and transfer it in a new tray after the last wash.

A tenfold gel volume of Soln. C is added. After 15 min, the silver solution is poured off and inactivated. The gel is washed three times with ddH_2O for 5 min each, and then it is developed with Soln. D as described in Protocol 2.4.2.1.

References

Tsai C-M, Frasch CE (1982) Anal Biochem 119:115

2.4.2.6 Enhancement of Silver Staining

Solutions/Reagents

- A 5% $FeCl_3$ (w/v) in ddH_2O
- B 3% oxalic acid (w/v) in ddH_2O
- C 3.5% $K_3[Fe(CN)_6]$ in ddH_2O
- D A, B, C, and ddH_2O are mixed immediately before use in a ratio of 1:1:1:7

After silver staining, the gel is thoroughly washed by several changes of ddH_2O (at least four changes of ten gel volumes, 10 min each). Then the gel is slightly agitated in Soln. D for 0.5 to 3 min and is washed again with ddH_2O. During this manipulation the color of the brown to black bands shift to blue, which is mostly better to recognize. Vanishing of different colors of individual bands is a disadvantage of this procedure.

References

Poehling HM, Neuhoff V (1981) Electrophoresis 2:141
Berson G (1983) Anal Biochem 134:230
Heukeshoven J, Dernick R (1985) Electrophoresis 6:103

2.4.2.7 Reducing of Silver-Stained Gels

Reducing (weakening) of overdyed gels should be the ultima ratio, because it is accompanied by a loss of information. Because the principle of silver staining is the same as in (silver halogenide) photography, reduction of dark background of a gel or to dark bands by oxidation of colloidal silver is done with a photographic reducer.

Solutions/Reagents

A 15% acidic fixing salt ($Na_2S_2O_3 \cdot 5H_2O$) (w/v), 1.2% thiourea (w/v) in ddH_2O
B 5% $K_3[Fe(CN)_6]$ (w/v) in ddH_2O

Solutions A and B are mixed immediately before use in a ratio given in Table 2.15.

The weakening or destaining is performed under visual control and finished by incubation with ddH_2O.

After extensive washings with ddH_2O (gels with exception of remaining bands and washings are absolutely colorless), the silver staining can be repeated or a further identification method is used.

Table 2.15. FARMER's reducer

Reducing power	Volume parts		
	A	B	H_2O
Fast and intensive	1	1	2
Medium	1	1	4

2.4.3 Copper Staining of SDS-PAGE Gels

Solution/Reagent

A 4% $CuCl_2$ (w/v) in ddH_2O

After electrophoresis, the gel is submerged without fixing into Soln. A (about 10 ml per milliliter of gel) for 5 min with gentle agitation. The gel is washed with ddH_2O for 2–3 min. The protein bands appear clear, whereas the rest is milky opalescent. This effect is best visible using a black background.

The authors describe the same detection limit if compared with Coomassie, but in our opinion, copper staining is less sensitive and has the main disadvantage that it is impossible to dry copper-stained gels. On the other hand, proteins are not denaturated by fixation and staining and can be eluted from gels in high yield.

References

Lee C, Levin A, Branton D (1987) Anal Biochem 166:308

Variant

A further option to stain by copper is the application of copper(II) phthalocyanine 3,4′,4″,4‴-tetrasulfonic acid(CPTS).

Solutions/Reagents

B 0.05 g copper(II) phthalocyanine-3,4′,4″,4‴-tetrasulfonic acid, tetrasodium salt, in 100 ml 12 mM HCl
C 12 mM HCl

Stain the gel for 1 h in Soln. B, pour off the liquid and destain by several changes of Soln. C until the background is colorless.

References

Bickar D, Reid PD (1992) Anal Biochem 203:109

2.4.4 Staining of Glycoproteins and Polysaccharides in Gels

2.4.4.1 Staining with Schiff's Reagent (PAS Staining)

PAS staining (periodic acid – Schiff's reagent) colors compounds with vicinal hydroxyl groups, i.e., mainly oligosaccharide side chains in glycoproteins, glycolipids, and nucleic acids. The sensitivity and stability is much lower than the silver staining Protocol 2.4.2.5 but more specific.

Solutions/Reagents

A 30% methanol (v/v), 10% acetic acid (v/v) in ddH_2O
B 7.5% acetic acid (v/v) in ddH_2O
C 1% sodium metaperiodate (w/v) in Soln. B
D Schiff's reagent: 1% basic fuchsin (magenta, rosaniline; C.I. 42510) (w/v), 1.9% sodium bisulfite in 0.15 N HCl. The solution is stored overnight in a refrigerator, and then a spatula of charcoal is added. After shortly stirring, the charcoal is filtered off. The colorless filtrate is stable for 4 weeks in a refrigerator.
E 1% sodium bisulfite (w/v) in 0.1 N HCl
F 0.01% 8-anilino-1-naphthalenesulfonic acid (ANS), ammonium salt, (w/v) in E

Shake the PAGE gel three times in Soln. A to remove SDS for 1 h. Then it is oxidized with Soln. C for 1 h at 4 °C, followed by at least four washings with Soln. B.

Incubate the gel in the dark with Soln. D for 1–1.5 h. Remove excess of SCHIFF's reagent by extensive washes with Soln. E. If the SCHIFF's reagent is not removed completely, the gel becomes purple during drying.

To visualize proteins in the presence of the red PAS staining, incubate the gel after finishing the PAS staining with Soln. F for 1–2 h. Excess of ANS is removed by washing with Soln. E.

The protein bands fluoresce if illuminated with UV.

Second staining with ANS

When no colored molar mass markers of special glycoprotein markers are used, the ANS staining is recommended, because the geometry of the gel is altered during the manipulations and comparison with Coomassie or otherwise stained gels is difficult.

References

Glossmann H, Neville DM Jr (1971) J Biol Chem 246:6339

2.4.4.2 Staining with Thymol

The sensitivity of staining of oligosaccharides with thymol is similar to the PAS staining, but not so time-consuming.

Solutions/Reagents

A 25% isopropanol (v/v), 10% acetic acid (v/v) in ddH_2O
B 0.2% (w/v) thymol (2-isopropyl-5-methylphenol) in Soln. A
C 80% (v/v) concentrated sulfuric acid, 20% (v/v) ethanol.

Caution! Corrosive!

D 30% methanol (v/v), 10% acetic acid (v/v) in H_2O

The SDS-PAGE gel is fixed by gentle agitation in Soln. A for 1–2 h. The washing is repeated with Soln. A three times, if possible one time overnight, to remove SDS, Tris, and glycine completely.

During the subsequent incubation with Soln. B for 90 min the gel becomes opalescent. After pouring of Soln. B, the gel is submerged in Soln. C. The gel becomes clear and the glycoconjugates appear as magenta bands. The color is stable for only a few hours.

After equilibration in Soln. D, the gel can be stained, for example, with Coomassie.

References

Gander JE (1984) Meth Enzymol 104:447
Gerard C (1990) Meth Enzymol 182:529

2.4.5 Staining of Blotted Proteins on Membranes

2.4.5.1 Staining on Nitrocellulose with Dyes

A fast and simple, but not sensitive, reversible staining is done with Ponceau S.

Important! *Not to be used with PVDF membranes!*

Staining with Fast Green is more sensitive, but not so contrastive.

Solutions/Reagents

A 0.2% Ponceau S (C.I. 27195) (w/v), 3% TCA (w/v) in ddH_2O
B 0.1% (w/v) Fast Green FCF, 1% acetic acid (v/v) in ddH_2O. Solutions A and B are reusable for several times.
C 0.05% (w/v) copper(II) phthalocyanine-3,4′,4″,4‴′-tetrasulfonic acid, tetrasodium salt (CPTS) in D
D ***destaining solution***: 1% acetic acid (v/v) in ddH_2O
E 12 mM HCl
F 0.5 M $NaHCO_3$in ddH_2O
G 20% ethanol (v/v) in H_2O

Ponceau S Staining

Reversible staining

Immediately after electrotransfer the nitrocellulose (NC) sheet is submerged into Soln. A by agitation for about 5 min. Then the sheet is rinsed with water or PBS until the bands appear and the background becomes white. The NC may be dried, but the color bleaches over a longer period. The sensitivity is about 100 ng/band.

If the stained NC is blocked with a protein-containing solution, the staining disappears completely. Detection of proteins by antibodies (immunoblotting) is not influenced by the staining with Ponceau S.

Fast Green Staining

A nitrocellulose sheet is stained with Soln. B immediately after electrotransfer or dot blotting. Soln. D is used for destaining. The detection limit is about 20 ng/band.

CPTS Staining

Staining or NC of PVDF sheets with CPTS is also reversible.

The NC membrane is agitated for 0.5 min in Soln. C after electrotransfer. Surplus of dye is removed by Soln. E. The detection limit for the blue-green bands is down to 10 ng/band.

For complete destaining the NC sheet is agitated in Soln. F for 15–30 min, followed by rinsing with Soln. G to remove traces of dye.

PVDF membranes are only destained by Soln. G.

References

Bickar D, Reid PD (1992) Anal Biochem 203:109

2.4.5.2 Staining on Nitrocellulose with Colloidal Gold

Solutions/Reagents

A 0.3% Tween 20 (w/v) in PBS
B 0.1% Tween 20 (w/v), 0.02% Carbowax 20 M (PEG 20000) in 10 mM citrate buffer, pH 3.0
C colloidal gold, diluted 1:1 with Soln. B (preparation of colloidal gold see Protocol 4.1.11.2)

After blotting, the nitrocellulose sheet is slightly moved in Soln. A for 15 min. Before incubation with Soln. B for 5 min, the sheet is rinsed with water. Soln. B is substituted by Soln. C and incubation is done for several hours or overnight with gentle agitation. Then the sheet is carefully rinsed with ddH_2O and dried. Proteins are visible as reddish bands.

References

Moeremans W, Daneels G, De Mey J (1985) Anal Biochem 145:315

2.4.5.3 Staining on PVDF Blotting Membranes with Dyes

Proteins blotted on PVDF membranes are stainable with Coomassie Brilliant Blue R250, but a relative intense background remains, which does not influence, for example, amino acid sequence analysis.

Solutions/Reagents

A 0.1% Coomassie Brilliant Blue R250, 40% Methanol (v/v), 1% acetic acid (v/v) in ddH_2O
B 80% methanol (v/v) in H_2O
C 0.005% Sulforhodamin B (Xylylene Red B, Acid Red 52, C.I. 45100) (w/v), 30% methanol (v/v), 0.2% acetic acid (v/v) in H_2O
Methanol

Staining with Coomassie Brilliant Blue R250

The membrane is rinsed with H_2O after protein transfer (electrotransfer or dot blot) and dipped to methanol for some seconds. Immediately after this (avoid drying), the membrane is agitated in Soln. A for maximum 2 min and destained by several changes of Soln. B. When the membrane is dry, blue bands are visible on a slight blue background. Membranes stained by this method are not suitable for any immunochemical detection.

Staining with Sulforhodamin B

After blotting, the membrane is washed with ddH_2O two times 10 min to remove salts. The membrane is dried overnight or by vacuum at RT. The dry membrane is incubated in Soln. C for 1–2 min, rinsed with ddH_2O, and dried.

CPTS is also suitable for staining of proteins on PVDF membranes (see Protocol 2.4.5.1).

References

Hancock K, Tsang VCW (1983) Anal Biochem 133:157

2.5 Electroelution from Gels

2.5.1 Preparative Electroelution of Proteins from Polyacrylamide Gels

This protocol is designed for proteins separated by SDS-containing PAGE systems. When proteins shall be eluted from gels of other systems, the respective electrode buffer has to be used instead of Soln. A and the polarity should be taken into account.

The content of glycerol or sucrose in Soln. A complicates the removal of SDS; therefore, the sample has to be dialyzed after electroelution. To avoid difficulties use Soln. A without glycerol or sucrose.

Solutions/Reagents

- A 25 mM Tris, 0.2 M glycine, 0.1% SDS (w/v), pH 8.3 [the solution may contain glycerol or sucrose up to 30% (w/v)]
- B 20 mM ammonium hydrogencarbonate buffer, pH 7.5; contains maximal 0.05% SDS if necessary
- C 80% TCA (w/v) in ddH_2O
- D 1 mM HCl in acetone

Acetone

The gel slices containing the protein of interest are minced in the presence of Soln. A (buffer volume nearly the same of gel volume), filled into a vessel of appropriate size and sonicated for 30–60 min.

Use an electrophoresis tube with an inserted sintered glass or polymer disk about 1 mm above one end and fill it at this side with buffer. Cover with a dialysis membrane with appropriate M_r cut-off, and fix with a rubber ring[19].

Fill the gel particles together with the surrounding buffer into the tube. The tube is inserted into the vertical electrophoresis apparatus (dialysis membrane down to anode), electrode buffer is poured, and electroelution is started with 10 mA per tube for 3–8 h, depending on the molar mass of the protein.

The eluted protein is taken off by a syringe from the space between sintered glass and dialysis membrane.

If the eluted protein is used for amino acid analysis, the sample has to dialyze three times 1 h each against 100-fold of Soln. B. After lyophilization, the protein is analyzed. SDS, which is not dialyzable under the given conditions, does not influence the determination.

If the protein concentration of the eluate is too low for further investigation, concentrate the sample using a centrifuge membrane concentrator (Centricon) with molar mass cut-off significantly lower than the M_r of the analyzed protein or by precipitation with TCA or acetone.

---

[19] Electro-eluter are commercially available. Their only limit is fixed volumes for gel and eluate.

TCA precipitation

To the sample containing at least 100 µg of protein per milliliter add one-seventh of Soln. C to a final concentration of 10% TCA. Incubate on ice for 30 min and centrifuge. Remove the supernatant (save it in case the protein did not precipitate) and wash the pellet twice with 100 µl ice-cold acetone each. The sample is dried in air.

Acetone precipitation

Dialyze the sample against three changes of 100 vol. of Soln. B. Lyophilize the sample, for example, in a SpeedVac, and then add 50 µl of ddH_2O. Add 450 µl of Soln. D and incubate at −20 °C for 3 h. Centrifuge, remove supernatant and wash twice with ice-cold acetone. The air-dried sample is subjected to further investigation.

References

Braats JA, McIntire KR (1978) In: Catsimpoolas N (ed.) Electrophoresis '78, Elsevier, North Holland, New York

Stone KL, Williams KR (1993) In: Matsudaira P (ed.) A practical guide to protein and peptide purification for microsequencing, 2nd edn. Academic Press, San Diego, p 48

2.5.2 Removal of SDS

Since SDS is practically not dialyzable, it must be removed by other techniques: extraction, precipitation, electroelution, and washing out from immobilized proteins.

Removal by extraction

A anhydrous acetone[20], triethylamine, acetic acid, ddH_2O 85:5:5:5 (v/v/v/v)

For extraction Henderson et al. gave the following procedure.

The sample is lyophilized after electroelution. To the dry sample add 500 µl of Soln. A. After vortexing the mixture is incubated in an ice bath at least for 1 h. Then the sample is centrifuged in a refrigerated centrifuge with minimal 3000 × g for 10 min. The supernatant is removed and the remaining pellet (sometimes not visible) is extracted twice with Soln. A as described. The remaining air-dried precipitate is free from SDS.

Removal by electroelution

The SDS-containing solution is supplemented with an excess of a nonionic detergent, e.g., Triton X-100. The electrode buffer is a buffer with low ionic strength, **without** SDS, and pH nearby the pI of the protein of interest. During electroelution SDS migrates to the anode.

[20] Anhydrous acetone is essential, since a water content above 5% decreases the protein yield by precipitation.

Removal by precipitation

Addition of potassium salts precipitates SDS nearly quantitatively, but there is the risk of co-precipitation proteins.

References

Henderson LE, Oroszlan S, Konigsberg W (1979) Anal Biochem 93:153
Hjelmeland ML (1990) Meth Enzymol 182:277

2.5.3 Electrotransfer of Proteins onto Membranes (Electroblotting; Western Blot): Semi-dry Blotting

The electrotransfer of proteins onto (non-specific) binding membranous sheets is named Western blot in contrast to the transfer of DNA (Southern blot) and of RNA (Northern blot). The main advantage of blotting procedures lies in the immobilization and presentation of macromolecules on the surface of a solid planar material. This presentation leads to an easy access of reactants in the opposite to the diffusion-controlled motion of reaction partners within gels or macroporous spheres.

Whereas gels are mainly electroblotted immediately after electrophoresis, it is possible to blot Coomassie or Amido Black stained gel too, but with lower efficiency and after soaking in ddH_2O for 15 min and an equilibration step in transfer buffer C (see below) for 30 min.

Blotting supports

Synthetic or semi-synthetic porous sheets, so-called membranes, able to bind proteins and other macromolecules non-specifically and with high capacity (amount of bound macromolecules per area), are mostly used as materials for planar blotting supports. The most applied material for protein blotting are nitrocellulose (NC) and polyvinylidendifluoride (PVDF) and with some less extended neutral or positively charged nylon.

Proteins interact with the membrane (support) by hydrophobic and charge-transfer forces and hydrogen bridges. The extent of these interactions depends on the accessibility of respective area of a protein. The accessibility is influenced, among other things, by the composition of the surrounding buffer, e.g., pH, ionic strength and/or chaotropic additives.

Two technical variants of electrotransfer from PAGE gels to membranes have been developed: the tank blotting developed by Towbin and the semi-dry blotting proposed by Kyhse-Anderson. With respect to transfer performance both systems are similar but the semi-dry system is easier to handle and less material-consuming; therefore, we followed the Kyhse-Anderson protocol for blotting from SDS PAGE gels in our lab.

The semi-dry blotting apparatus consists of two plates (anode and cathode) made from graphite, glassy carbon, or stainless steel and a sandwich consisting of buffer-soaked filter paper, transfer membrane and electophoresis gel. The sandwich is made from several layers of filter paper (three to four sheets of, for example, Whatman MM3), the membrane, and the gel (Fig. 2.3). Filter paper and membrane should have the same size as the gel.

***Important!** Never touch gels or membranes with fingers. Wear gloves and/or use tweezers!*

Solutions/ Reagents[21]

A ***first anode buffer***: 300 mM Tris in ddH_2O (do not correct pH)

B ***second anode buffer***: 25 mM Tris in ddH_2O (do not correct pH)

C ***cathode buffer***: 25 mM Tris, 40 mM ε-aminocaproic acid (EAC, 6-aminohexanoic acid), 0.04% SDS

If smaller proteins shall be blotted, add up to 20% methanol to the buffers A, B, and C.

The formation of the transfer unit starts with the anodic filter paper stack. Two to three sheets of filter paper are soaked in buffer A and placed onto the anode. Air bubbles are removed using

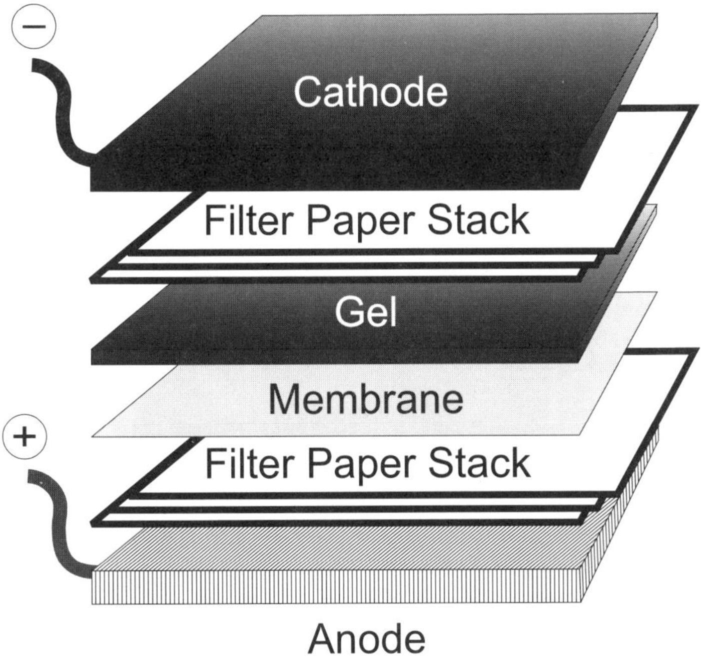

Fig. 2.3. Principle of semi-dry blotting

[21] Kondo et al. simplified the buffers. They used a buffer of 25 mM Tris, 192 mM glycine for the anodic side, and 25 mM Tris, 192 mM glycine, 0.1% SDS as cathode buffer. We did not find significant differences between both buffer systems. (Kondo M, Harada H, Sunada S, Yamaguchi T (1991) Electrophoresis 12:685)

a glass rod. Then two to three sheets follow soaked with Soln. B. The membrane is wetted with Soln. B and placed on top of the filter paper. When using the very hydrophobic PVDF membranes, submerge the membrane in methanol (take care that the whole membrane is wetted, i.e., the color changes from white to opalescent) for some seconds and transfer it immediately to Soln. B. Do not leave the wet PVDF membrane on air, because it dries very quickly.

The SDS-PAGE gel is agitated for few minutes in Soln. B and then put onto the membrane. Some layers of filter paper soaked in Soln. C finish the sandwich. The blotting apparatus is covered with the cathode plate and the electrodes are plugged into the power supply. Electrotransfer is performed with a constant current of 0.8 mA/cm^2 of gel area for 0.5–2 h at RT. The voltage does not exceed 15 V.

Multiple electrotransfer of up to three gels is possible. On top of the filter papers soaked with Soln. B a wetted dialysis membrane is placed and then the sandwich is repeated. Transfer conditions are similar to the single sandwich except the total gel area taken into account for calculation of applied current.

Buffer modifications

If the conditions of the receiving material (e.g., activated paper instead of nitrocellulose) forbid the use of Tris-containing buffers, the following substitutes may be used:

A′: 0.3 M triethanolamine (M_r 149.2), pH 9.0; B′: 25 mM triethanolamine, pH 9.0; C′: 25 mM triethanolamine, 40 mM butyric acid, pH 9.0 to 9.2.

Important! *Use ultrapure triethanolamine, free of primary amines!*

Proteins blotted on PVDF can be subjected directly to sequence analysis. Specialized protocols for this purpose are referred to, for example, by Jungblut (1997).

References

Beisiegel U (1986) Electrophoresis 7:1
Jungblut P (1997) In Kamp RM, Choli-Papadopoulou T, Wittmann-Liebold B (eds.) Protein structure analysis. Springer, Berlin, p. 215
Kyhse-Andersen J (1984) J Biochem Biophys Meth 10:203
Towbin H, Gordon J (1984) J Immunol Meth 72:313

2.5.4 Immunochemical Detection of Antigens After Electrotransfer (Immunoblotting)

Macromolecules blotted onto membranes are detected very specifically and sensitively by antibodies. The prerequisite is the existence of epitope(s) on the surface of blotted antigens after denaturation during SDS-PAGE. Very often, even fragile epitopes (which are in each case three-dimensional structures formed by the macromolecule's constituents) reconstitute during washing and blocking of the blotting membrane, but sometimes the lack of an antibody reaction is not due to the missing target macromolecule: the required epitopes are not presented under the distinct conditions.

Immunochemical detection is possible in one step, if the detecting antibody carries the signal-forming principle. But also cascades of steps are used to identify the first-bound antibody and by it the antigen immobilized by blotting. Antibodies form these cascades, i.e., if the first bound antibody is from species A, e.g., rabbit, a second antibody from other species, e.g., goat, directed against the primary, is used.

Biotin/(strept)avidin

A second possibility is the modification (conjugation) of an antibody by a label, e.g., biotin, which is detected later on by a specific receptor, e.g., (strept)avidin. In each case the last part of such a cascade has to carry a measurable label. Such labels are enzymes, fluorescent dyes, colloids, radioactive isotopes, paramagnetic substances, and others.

As indicator enzymes horseradish peroxidase (HRP or HRPO), alkaline phosphatase (AP), or β-galactosidase, are favored, since they are relatively robust, have a high product-forming rate, are easy to purify, and are cheap. The most used colloids are from gold, silver, and iron, and iodine isotopes are mostly taken as radioactive labels in immunoassays.

The first step in immunochemical detection of proteins after electrotransfer is blocking the support with an inert material to inactivate further non-specific binding of protein. The blocking reagent should cover the membranes at those areas where no blotted protein is bound and should not react with any of the reactants of immunochemical detection cascade as indicated by no non-specific staining, i.e., resulting in blank background of the membrane.

The following protocol is an example for blocking, working very well in numerous cases, but optimization by use of other blocking reagents is worth checking every time.

Solutions/Reagents

A 0.05% Tween 20 (w/v), 0.1% serum albumin[22] (w/v) in PBS
B 0.05% Tween 20 (w/v) in PBS

After electrotransfer the wet membrane is rocked in about 1 ml of Soln. A per square centimeter of membrane at RT three times for 10 min each[23].

[22] Mostly bovine serum albumin (BSA) is used. Gelatin from cold fish, inactivated calf serum, casein hydrolysate, non-fat dry milk, polyvinylpyrrolidone, or Tween 20 are also suitable. Concentrations from 0.1 to 0.5% (w/v) are sufficient. Casein and milk are not of first choice if (strept)avidin conjugates are used.

[23] If anti-peptide antibodies are used the detection sensitivity is often increased if the humid membrane is treated in a steam autoclave at 120 °C for 30 min before blocking (SWERDLOW PS, FINLAY D, VARSHAVSKY A

When blocking is finished, blots can be dried on air, are stored frozen, or are incubated with the proper antibody dilution in Soln. A. If antisera are used, the dilution should be 1:100 at least to prevent non-specific adsorptions. Purified polyclonal or monoclonal antibodies are mostly diluted much higher.

0.2 ml/cm^2 of antibody dilution is sufficient in most cases. To save antibodies, put a 2 × 80 mm blotting strip into a centrifuge tube, add 300 µl of antibody dilution, close the tube, and incubate on a roller shaker.

Incubation of blotted proteins with antibody dilutions should not last less than 30 min and not more than 2 h at RT or may be kept in the refrigerator overnight.

Use of pre-immune serum

Especially if antisera are used, an incubation of a parallel blotting strip with a non-specific control serum from the same species (better: pre-immune serum from the same animal) with the same dilution ratio should be done to exclude false-positive results. These false-positive results for proteins with higher molar mass are often found if rabbit sera are used.

After primary antibody incubation, the blot is washed three times with Soln. B for 5 min each. If the primary antibody carries a label (e.g., enzyme or radioactive isotope), the detection reaction follows this washing.

If a second species-specific and labeled (conjugated) antibody or a specific receptor, e.g., ^{125}I-Protein A or HRP-conjugated streptavidin, is used, the blot is incubated with an appropriate dilution at RT for 30 min and washed again three to four times with Soln. B.

The dilution of primary antibody as well as for secondary antibody (conjugate) is determined empirically, but for guidance the following dilutions of antibodies or conjugates are practicable:

Serum	1:50–1:10 000
Purified antibodies	1:500–1:1 000 000
Species-specific antibodies and their enzyme conjugates	1:1000–1:50 000

2.5.4.1 Detection Using Horseradish Peroxidase (HRP)

Horseradish peroxidase catalyzes the cleavage of hydroperoxide substrates forming active oxygen, which oxidizes molecules resulting in a colored product. For application in Western blots the reaction product must be insoluble in aqueous buffer solutions.

Solutions/Reagents

A 50 mg/ml 4-chloro-1-naphthol (4-CN), 3,3′,4,4′-tetraaminodiphenyl ether, or 3,3′-diaminobenzidine (DAB) in DMF (stock solution)

***Caution:** DAB is potentially carcinogenic!*

(1986) Anal Biochem 156:147). If antibodies are used raised against whole proteins, this thermal denaturation mostly destroys the epitopes on blotted proteins.

B 600 µl 0.1% H_2O_2-urea adduct (w/v) in ddH_2O, 10 µl 10% $CuSO_4$ (w/v), 5 µl 10% $NiSO_4$ (w/v) or $NiCl_2$ and 9.35 ml PBS are mixed, then 40 µl of Soln. A are added. Prepare reagent immediately before use.
C 2.4 mg/ml (20 mM) 3,3′,5,5′-tetramethylbenzidine (TMB) in DMSO (stock solution).
D 0.2 M sodium acetate pH 5.0.
E 20 mM ammonium molybdate ($(NH_4)_6Mo_7O_{24} \cdot 4H_2O$, MW 1235.86) in dd$H_2O$. Stable at RT.
F 0.1% H_2O_2-urea adduct (10.6 mM H_2O_2) in ddH_2O. Stable for several days in a refrigerator.
G ***precipitating TMB***: mix Soln. C, D, E, F, and ddH_2O in a ratio of 0.5:5:1:2:1.5 prior to use[24]
H 0.1 M H_2SO_4

4-Chloro-1-naphthol (4-CN) and diaminobenzidine (DAB) are used in Soln. B; staining with tetramethylbenzidine (TMB) is done in Soln. G at RT for 10–20 min.

After color development Soln. B and G, respectively, are poured off and the membrane is rinsed intensively with water.

When precipitating TMB is used, a visual intensification is possible by shaking the membrane in Soln. H for some seconds. The blue color of bands switches to brown and the background becomes paler.

Stained blots are dried on air on a sheet of filter paper. If radioactive compounds are also transferred, autoradiography of the dry membrane is possible at −70 °C.

2.5.4.2 Detection Using Alkaline Phosphatase (AP)

Solutions/Reagents

A 100 mM Tris, 100 mM NaCl, 5 mM $MgCl_2$, pH 9.5
B 37.5 µl of 5% 5-bromo-4-chloro-3-indolyl phosphate (X-phosphate, BCIP) (w/v) in DMF and 50 µl of 7.7% nitroblue tetrazolium chloride, toulidinium salt (NBT) (w/v) in 70% DMF (v/v), added and mixed immediately before use

Alkaline phosphatase cleaves phosphate esters at medium alkaline pH in the presence of magnesium ions.

When AP will be taken for detection in Western blots, after antibody processing the membrane is equilibrated twice for 5 min each in Soln. A. Equilibration buffer is poured off and the color is developed in Soln. B completely covering the membrane. During this step the membrane should be moved very gently only.

Visualization Soln. B is poured off after 5–15 min, and then the membrane is washed for some seconds with water and dried on air covered by filter paper. Since the dye is air- and light sensitive and bleaches over a longer period, the dry membrane should be wrapped with a foil and stored protected from light.

[24] Precipitating TMB ready to use is offered by several companies.

2.5.5 Chemiluminescence Detection on Blotting Membranes

AP as well as HRP as part of a ligand/ligate (antigen/antibody) complex is detected very sensitively by chemiluminescence.

Substrates for AP are, e.g., AMPPD (disodium 3-(4-methoxyspiro{1,2-dioxethane-3,2′-ticyclo[3.3.1.1^{3,7}]decan}-4-yl)phenyl phosphate) or CSPD (disodium 3-(4-methoxyspiro{1,2-dioxethane-3,2′-(5″-chloro)tricyclo[3.3.1.1.^{3,7}]decan}-4-yl)-phenyl phosphate).

HRP splits hydroperoxides and resulting oxygen reacts with luminol, which gives chemiluminescence. Important: The different substrates follow different kinetics, i.e., chemiluminescence of luminol starts very quickly and also declines soon, and AP-produced chemiluminescence raises and disappears slowly.

Reagent kits are commercially available for both enzymes. These kits are optimized to achieve sensitivities comparable to those of autoradiography. Disadvantages are more sophisticated handling and higher expenditure for documentation. For some optimized chemiluminescence kits the suppliers suggest a special blocking step for increasing sensitivity. Consult instructions of reagent manufacturer for detailed information.

After exposure to an X-ray film or photography in the dark room, the blot can be rinsed with TBS and used for a development with a chromogenic substrate.

2.5.5.1 Chemiluminescence Using HRP

(GE Healthcare; formerly Amersham Biosciences ECL Kit Protocol)

Solutions/Reagents

A 0.1% Tween 20 (w/v) in PBS

After incubation with antibody and second antibody HRP-conjugate, wash the blot three times with about 1 ml/cm^2 of Soln. A for 5 min each. Be sure that no buffer used after the incubation with conjugate contains sodium azide or another inhibitor of HRP.

Pour off the washing buffer and add ECL reagent mix (1 vol. reagent 1 and 1 vol. reagent 2, about 0.5 ml/cm^2) and incubate for 30 s to 1 min at RT. Remove membrane, drain, and place face down to a thin clear plastic wrap. Form a moisture-tight envelope from the remaining wrap and place the wrapped membrane with blot face down onto an X-ray film in a dark room. Be sure that no liquid contacts the film.

The exposure time ranges from 2 to 30 min. An alternative to X-ray film is a digital camera.

2.5.5.2 Chemiluminescence Using AP

(Roche protocol)

Solutions/Reagents

A 0.1% Tween 20 (w/v) in TBS

B 0.1 M Tris, pH 9.5, 0.1 M NaCl, 5 mM $MgCl_2$ (TNM buffer)

C AMPPD stock solution (from the supplier)

Wash the blot after the antibody and conjugation incubation steps with Soln. A, followed by an equilibration in 2 ml/cm^2 of Soln. B for 2–3 min.

Dilute Soln. C 1:100 with Soln. B.

Substitute Soln. B by the AMPPD dilution (0.2 ml/cm^2) and incubate for 5 min. Drain the blot and wrap it with a solvent-tight foil. Incubate the wrapped membrane at 37 °C for 5 min, and then expose the blot face towards to the film in an X-ray film cassette for 15 min to 24 h or photograph it with a digital camera.

It is possible to make a first exposure for, for example, 15 min, followed by a second and/or third with longer exposing times on new films each.

2.5.6 Carbohydrate-Specific Glycoprotein Detection After Electrotransfer

Specific detection of oligosaccharides is possible using lectins, carbohydrate-binding proteins from very different sources. A se-

Table 2.16. Selected lectins

Lectin[a]	Abbre-viation	MW (kD)[b]	Specific carbohydrate
Aleuria aurantia lectin	AAA		Fuc(α1-6)GlcNac[c]
Concanavalin A	Con A	102	α-D-Man, α-D-Glc
*Datura stramonium*lectin	DSA	86	Gal(β1-4)GlcNAc
Galanthus nivalis lectin	GNA	52	Man(α1-3)Man
Gorse lectin	UEA I	68	α-Fuc
Gorse lectin	UEA II	68	$(GlcNAc)_2$
Lentil lectin	LcL	49	α-D-Man, α-D-Glc
Maackia amurensis lectin	MAA	130	NeuNAc(α2-3)Gal
Mistletoe lectin	VaL	115	β-D-Gal
Peanut lectin	PNA	120	β-D-Gal_{1-3}-D-GalNAc
Phytohemagglutinin	PHA-E	128	(non-specific)
Phytohemagglutinin	PHA-L	126	(non-specific)
Phytolacca americana lectin	PWM	32	GlcNAc(β1-4)$GlcNAc_{1-5}$
Ricinus lectin I	RCA_{120}	120	β-D-Gal
Ricinus lectin II	RCA_{60}	60	D-GalNAc
Sambucus nigra lectin	SNA	140	NeuNAc(α2-6)Gal
Snail lectin	HPA	79	α-GlcNAc
Soybean lectin	SBA	110	α-GalNAc
Wheat germ agglutinin	WGA	36	$[D\text{-}GlcNAc]_n$, NeuNAc

[a] The terms "lectin" and "agglutinin" are used synonymously
[b] Molar mass under physiologic conditions
[c] *Fuc* fucose, *Gal* galactose, *GalNAc* N-acetylgalactosamine, *Glc* glucose, *GlcNAc* N-acetylglucosamine, *Man* mannose, *NeuNAc* N-acetylneuraminic acid (sialic acid, NANA)

lection of lectins and their specificities is given in Table 2.16. One should keep in mind that most of the lectins are specific for oligosaccharides; monosaccharides compete mostly in higher concentrations (CUNNINGS 1994).

For checking the specificity of interactions between glycoprotein and lectin, an incubation in the presence and the absence of 1–5 mM of the respective monosaccharide, e.g., α-methylmannoside for ConA or GlcNAc for WGA, cf. Table 2.16), is recommended.

For identification the lectins must be labeled. The best labels are biotin (streptavidin-enzyme conjugate) or digoxigenin (anti-digoxigenin antibody-enzyme conjugate), but in the case of Concanavalin A, direct labeling with HRP is possible (see below).

Solutions/Reagents

A 40 mg/ml of protein (e.g., serum albumin) in 0.1 M sodium acetate, pH 4.5 are incubated with periodic acid (final concentration 10 mM) at RT for 6 h. Then add glycerol to 10 mM and dialyze twice against PBS for 2 h each (stock solution). For use dilute 1:20 with PBS.
or 0.5% polyvinylpyrrolidone (PVP) (w/v) in PBS
or 1% Tween 20 (w/v) in PBS (Attention! This concentration of Tween 20 sometimes removes proteins form the membrane!)
B 1 mM $CaCl_2$, 1 mM $MnCl_2$ in PBS

Block the membrane after electrotransfer with Soln. A for 30 min at RT. Wash three times with PBS.

Incubate with the biotin or digoxigenin-labeled lectin, diluted in Soln. A to 5–50 µg/ml, for 1 h at RT. Wash at least three times with PBS or TBS.

Incubate with streptavidin-HRP conjugate and anti-digoxigenin antibody conjugate for 15–30 min at RT. Wash again thoroughly.

Stain as described in Protocols 2.5.4 or 2.5.5.

If the lectin Concanavalin A shall be used, dilute it in Soln. B. Incubate as described above. After washing, incubate with 50 µg/ml HRP in Soln. B for 30 min, wash with Soln. B again, and stain as described in Protocols 2.5.4.1 or 2.5.5.1.

References

Glass WF II, Briggs RC, Hnilica LS (1981) Anal Biochem 115:219

2.5.7 General Carbohydrate Detection on Western Blots

Carbohydrates are easily oxidized by periodate to aldehydes which react with primary amines or hydrazines. If the formed hydrazide carries a specific ligand, e.g., biotin or digoxigenin, these ligands immobilized via the blotted macromolecule are very sensitively detected by the respective enzyme conjugates.

Solutions/Reagents

A PBS
B 0.1 M sodium acetate, pH 5.5
C 10 mM $NaIO_4$ in B
D TBS

Agitate the unblocked membrane after electrotransfer in Soln. A three times for 10 min. Equilibrate the membrane in Soln. B for 1 min, discard Soln. B and incubate with Soln. C (about 0.5 ml/cm^2) for 20 min. Wash three times with Soln. A.

Add a sufficient amount of the hydrazide (e.g., 0.2 µg/ml of digoxigenin hydrazide (digoxigenin-succinyl-ε-aminocaproic acid hydrazide)[25] or 25 µg/ml of Biotin-LC-Hydrazide[26]) and incubate for 1 h at RT. Wash three times for 10 min with Soln. D and block the membranes for 15 min with an appropriate blocking solution, e.g., 1% non-fat dry milk in Soln. D. Wash with Soln. D.

Incubate with the corresponding enzyme conjugate (e.g., anti-digoxigenin AP conjugate or streptavidin-HRP) for 30 min, wash carefully and perform the staining as described in Protocols 2.5.5 and 2.5.4.

References

O'Shannessy DJ, Voorstad PJ, Quarles RH (1987) Anal Biochem 163:204

2.5.8 Affinity Blotting

Covering the blotting membrane with a suitable capture ligate, e.g., antibody, increases the sensitivity for detecting the corresponding ligand, e.g., antigen.

Solutions/Reagents

A 0.1 M $NaHCO_3$, pH 8.8
B 0.01% glutaraldehyde (w/v) in ddH2O
C TBS
D 0.02% $NaBH_4$ (w/v) in TBS
E 0.5% Tween 20 (w/v) in TBS

Dilute the affinity-purified capture antibody to 10–100 µg/ml in Soln. A. Incubate a new piece of blotting membrane with the antibody dilution (0.2–0.5 ml/cm^2) at RT for 30 min.

To save antibody, use a sealable tube and a roller shaker. If such a tube is used, make sure that the receiving side of the membrane is directed to the lumen of the tube.

Wipe off remaining antibody dilution after incubation with filter paper and incubate the wet membrane in Soln. B for 30 min. Rinse with Soln. C and stop the fixation by incubation with Soln. D for 2 min. Block with Soln. E for 30 min, wash three times with Soln. C, and dry in vacuo.

[25] Roche Glycan Detection Kit
[26] Pierce Chemical Co.

The activated dry membrane can be stored in a refrigerator for several weeks.

Blot the proteins onto the affinity membrane either by dot blot (Protocol 4.8), by electrotransfer (Protocol 2.5.3), or by capillary blotting (handling analogously to Southern blotting Protocol 2.5.9).

Capillary blotting is recommended, if the gel is mounted onto a impermeable carrier. Put the affinity membrane with the coated face towards the gel onto the gel, cover the backside of the membrane with several layers of dry filter paper, wrap the stack into a water-tight foil, and store it under moderate pressure in a refrigerator for several hours or overnight.

Open the sandwich, remove filter paper and gel, wash the membrane several times with TBS or PBS, and incubate with a (labeled) antibody, generated in another species and directed against a second epitope. Wash again and detect the second antibody by the usual procedures.

2.5.9 Transfer of Nucleic Acids (Southern and Northern Blot)

The transfer protocols are the same for DNA and RNA. In opposition to the methods given in Protocol 2.5.3, the forces driving the biomolecules from the separating gel to the receiving membrane are diffusion and capillary flow. This type of transfer is also applicable for proteins, but because the pores of polyacrylamide gels used for protein separation are mostly smaller, transfer times are longer and transfer efficiencies are lower than by electrotransfer.

Transfer (blotting) of DNA to membranes was introduced by E.M. Southern in 1974, and as a pun the transfer of RNA as the opposite of DNA was named Northern blot and that of proteins got the direction West.

Solutions/Reagents

A 0.25 N hydrochloric acid
B 0.6 M NaCl in 0.2 N NaOH
C 1.5 M NaCl in 0.5 M Tris · HCl, pH 7.0
D 3 M NaCl in 0.3 M sodium citrate buffer, pH 8 ("20 × SSC")

Agitate the nondenaturated DNA gel (cf. Protocol 2.2.1) carefully in its tenfold volume of Soln. A at RT for 30 min. Skip this step if the DNA is significantly smaller than 5000 bp.

Decant the hydrochloric acid, add ten gel volumes of Soln. B, and shake for additional 30 min, repeating this step with Soln. C instead of Soln. B.

Place a glass plate somewhat larger than the gel over a tray filled with Soln. D. Wet thick filter paper, e.g., Whatman 3MM, with Soln. D and spread it over the glass plate, the ends dipping into Soln. D. Now place the gel on the filter paper and remove air bubbles. Cover the sides of the gel by strips of plastic wrap and put the wetted membrane onto the gel.

> ***Important!*** *Do not touch the membranes with unprotected fingers, use tweezers and gloves!*

Squeeze out air bubbles using a glass rod. Moisten the membrane with Soln. D and place five sheets of filter paper of gel size onto the membrane. Cut paper towels to the same size as the membrane and stack these on top of the filter paper to a height of about 4 cm. Lay a glass plate on top of the pyramid and place a weight on top to hold everything in place. Transfer at RT overnight.

Remove paper. Mark the position of the gel and the contact side between gel and membrane on the membrane with a pencil. Rinse the membrane with 1:10 diluted Soln. D and allow the membrane to dry on air. Cover the membrane on both sides with filter paper, fix filter paper and membrane with paper clips, and bake for 2–3 h at 80 °C.

The baked nictrocellulose is ready for hybridization experiments.

If other membranes are used instead of nitrocellulose, read their instructions for use.

References

Southern EM (1975) J Mol Biol 98:503

Sambrook J, Fritsch EF, Maniatis T (1989) Molecular cloning. A laboratory manual. 2nd. ed., vol. 2, Cold Spring Harbor Laboratory Press

Brown T (1999) In: Current protocols in molecular biology. Wiley, New York, Unit 2.9.1

2.6 Drying of Electrophoresis Gels

The best way for storage of polyacrylamide gels is drying in vacuo. Photograph or scan the gel before drying.

A lot of gel dryers are commercially available which allow to perform the drying process at elevated temperature resulting in shortened time.

Wet a sheet of filter paper a little bit larger than the gel with 5% glycerol in water. Place this sheet on the porous surface of the apparatus, lay the gel onto the paper, and cover it with wetted cellophane or another type of plastic foil.

> ***Important!*** **Cover only one side of the gel with a water-tight foil!**

Cover the gel with the rubber cloth, connect to the vacuum pump, and switch on the heating. When the temperature at the position of the gel is the same as that of the surrounding temperature, continue drying for about 30 min and then open the apparatus. If the vacuum is switched off too early, the gel becomes destroyed irreversibly. To prevent this cracking, especially of gels with high acrylamide concentration, supplement the last detaining solution with 1% glycerol.

Agarose gels and polyacrylamid gels with %T < 8 are dried on air without vacuum. The gel is wrapped in cellophane and the envelope is fixed in a frame. The drying needs up to several days.

Agarose gels with less than 1% agarose on a support (glass plate or GelFix foil) dry on air without any manipulations (cf. Protocol 4.8.4).

2.7 Autoradiography of Radioactive Labeled Compounds in Gels

> ***Important!*** *Pay attention to the local and state rules governing the use of radioactive materials!*

Autoradiography is the method of choice for detection and quantification of radioactive labeled samples in electrophoresis gels, blotting, or hybridization filters using X-ray films. The main drawback of X-ray films is a limited linear range of image density with respect to the amount of radioactivity.

Whereas samples containing strong radiation emitters as ^{32}P or ^{125}I are subjected to autoradiography directly, ^{3}H, ^{14}C, or ^{35}S labeled probes need an enhancement by solid-state scintillation.

Mostly dried gels or membranes are used in autoradiography. If wet gels have to be exposed to films, wrap the gel into an impermeable thin foil to prevent chemical development of the film.

Sensitivity of X-ray films can be increased by use of an intensifying screen or by pre-flashing. The following protocol describes autoradiography using intensifying screen and solid scintillation, respectively.

Autoradiography in the Presence of an Intensifying Screen

Open an X-ray film cassette in a dark room illuminated with a weak green or red light. Mount the intensifying screen, place a piece of X-ray film on it, and lay the dry gel or membrane on top. If no radioactive ink or radioactive labeled markers are used, prick a needle through gel and film to allow identification of the exact position of the gel after development.

Close the cassette and expose at −70 °C for several minutes to hours. Since the time of exposure depends on the amount of radioactivity, this time has to be proved empirically.

After exposure, open the cassette and develop the film according to manufacturer's instructions in the dark room. After fixing, wash the film for 15 min in running water and dry on air.

Pre-flashing increases the sensitivity of X-ray films. Pre-flash needs some experience to catch the right conditions for high sensitivity and low background.

Autoradiography in the Presence of a Scintillator

The radiation of ^{3}H, ^{14}C and ^{35}S has to be converted into light by scintillators.

Solutions/Reagents

- A methanol:glacial acetic acid:pure water 5:5:5 (v/v/v)
- B 16% sodium salicylate (w/v) in ddH_2O
- C radioactive ink: mix some kBq of any ^{14}C-labeled compound with a little amount of refill and refresh a pen with it. Handle this pen with care and avoid contaminations.[27]

Equilibrate the gel after staining in Soln. A for 20 min. Wash the gel with ddH_2O twice in a 20-fold of its volume for 20 min each, followed by an equilibration in Soln. B for 30 min. Dry the gel and expose to the X-ray film at −70 °C as described above.

Blotting membranes and TCL plates are processed analogously[28]. Incubate the dry plate or membrane in Soln. B for 5–10 min, dry again, and expose to film.

Variant

The scintillator 2,5-diphenyl-oxazole (PPO) in DMSO may be used instead of salicylate:

Solutions/Reagents

- D 22% PPO (w/v) in DMSO (solution is reusable for several times)

> ***Attention!*** *Wear gloves and avoid skin contact to DMSO-containing solutions, because DMSO penetrates skin very effective.*

Incubate the gel fixed with Soln. A in a 20-fold volume of DMSO for 30 min. Replace DMSO by the same volume of fresh DMSO and agitate gently for additional 30 min. The DMSO bathes may be used several times when stored separately and used in the same order.

Shake the gel in 4 vol. of Soln. D for 2–3 h and transfer to water after that time. Let the gel for 1 h in water and then dry.

Expose the dry gel to X-ray film in a cassette with intensifying screen at −70 °C as described above.

Quantification of radioactivity is possible by densitometry (scanning) of the developed film. It should be taken into consideration that the optical density of the film is not linear proportional to the amount of radioactivity, especially at lower radioactivity. Dot defined amounts of radioactivity onto a part of the gel or membrane and use the obtained darkening to construct a calibration curve.

Another possibility of quantification is to cut the bands of interest and counting in a liquid scintillation counter after mincing the gel pieces.

---

[27] Instead of radioactive ink, pens with phosphorescent dyes are available.

[28] LUCHER LA, LEGO T (1989) Anal Biochem 178:327

References

Chamberlain JP (1979) Anal Biochem 98:132

Laskey RA (2002) Efficient detection of biomolecules by autoradiography, fluorography or chemiluminiscence. Review 23, Amersham Biosciences

Slater RJ (ed) (1990) Radioisotopes in biology: a practical approach. Oxford Press, New York

Voytas D, Ning Ke N (1999) Current protocols in molecular biology. Wiley, New York, Appendix 3A

3 Chromatography

3.1 Thin-Layer Chromatography

Thin-layer chromatography (TLC) has passed its heyday as an analytical procedure and has been surpassed by more sensitive and automatic methods, but it still has some advantages in the analysis of complex lipids. Despite its decrease in usefulness, because of the simple realization and low-cost apparatus, some examples for TLC are given in this chapter.

3.1.1 Identification of the N-terminal Amino Acid in Polypeptides (TLC of Modified Amino Acids)

Prerequisite is the existence of a primary amino group at the N-terminal end of a polypeptide, i.e., chemical or posttranslational modifications of this amino group, e.g., by methylation or acetylation, prevents success. If the amino group is not protected or the amino acid chain is not branched, this method suits well for examination of the uniformity of a purified protein.

If spots of multiple labeled amino acids (e.g., lysine, cysteine, histidine, tyrosine) are found, check whether the intensity of the multiple labeled derivatives is in the same order as the mono-labeled derivative. Only in that case could the determined amino acid be N-terminal.

All chemicals and reagents must be of highest purity ("Sequence Grade"); especially water must be at least freshly double distilled.

Solutions/Reagents

A *coupling reagent*: 1.4 mg DABITC (4-(dimethylamino)azobenzene-4′-isothiocyanate, M_r 282.4, recrystallized from acetone) is dissolved in 1.0 ml acetone. The solution is divided into 40 µl aliquots, and acetone is evaporated by nitrogen. The aliquots are stable in a desiccator at room temperature for months. For reaction freshly dissolve 1 aliquot (corresponding 56 µg) in 20 µl of pyridine.
B PITC (phenylisothiocyanate), undiluted, Sequence Grade
C n-heptane/ethylacetate 2:1 (v/v)
D 40% TFA (trifluoroacetic acid) (v/v) in ddH_2O
E 50% pyridine (v/v) in ddH_2O
F *diethylurea marker*: mix 6 µl diethylamine and ethanolamine each with 100 µl Soln. D, flush with nitrogen and heat for 1 h

at 55 °C. Dry in vacuum, dissolve in 1 ml ethanol, aliquot to 100 µl portions and store at −20 °C
butylchloride or butylacetate
TFA, anhydrous
HCl, concentrated

Solvents for TLC

1 acetic acid/H_2O 1:2 (v/v)
2 toluene/n-hexane/acetic acid 2:1:1 (v/v/v)

This protocol gives a survey of the method. For details and special tips see CHOLI-PAPADOPOULOU et al. (1997).

Derivatization, cleavage, and conversion

Dry a desalted sample of the protein or polypeptide (0.5–5 nMoles) over solid KOH in vacuum. Add 20 µl of Soln. E and flush with nitrogen. Add the described aliquot of A in pyridine (20 µl). Flush the tube with nitrogen, and then close it and heat it to 52 °C for 20–30 min. Add 2 µl reagent B, flush again with nitrogen and incubate the closed tube for further 20 min at 52 °C.

Remove uncoupled reagents by extraction with Soln. C. For this purpose, add 200 µl of Soln. C to the reaction mixture, centrifuge, carefully withdraw the supernatant and repeat extraction twice. Discharge the supernatants and dry the aqueous phase in vacuum. To remove residual traces of water, add 10 µl of ethanol to the tube and dry again.

Add 20 µl of anhydrous TFA to the dry sample in a nitrogen atmosphere. The solution changes to red. Heat the closed tube to 55 °C for 10 min, and then blow off the TFA in a hood.

Wet the residue with 30 µl of H_2O and extract the solution twice with 30 µl butylchloride or n-butylacetate. Separate phases by centrifugation and collect the upper organic layer into a new tube.

Evaporate the organic solvent with nitrogen, and then add 10 µl of Soln. D. Flush the tube with nitrogen, close it, and incubate it for 30 min at 55 °C. Then dry the solution in vacuum.

Chromatography and identification of N-terminal amino acid

Dissolve the dry residue of the conversion step in 5 µl of ethanol. Spot 0.5–1 µl of this solution together with the same amount of marker Soln. E onto the start point of a 25 × 25 mm polyamide TLC sheet, about 3 mm distant from an edge.

Cover a chromatographic chamber, with at least 30 mm diameter, using filter paper and equilibrate with Solvent 1. After equilibration, pour off the solvent and fill up new solvent to a hight of 2 mm.

Develop the sheet for 4–5 min and remove it when the solvent front is about 2 mm from the top of the sheet. Remove the sheet and dry it with a cold fan.

Equilibrate a second chromatographic chamber with Solvent 2 as described above. The dried sheet is turned to 90° and developed for the second dimension until the solvent front reaches about 5 mm below the top.

Dry the sheet again and keep on the vapor of concentrated HCl. Identify the red spots in comparison with a chromatogram of DABTH amino acid test mixture and determine migration relative to the bluish spots of the markers (schematic drawing of a 2 × 2-cm TLC plate is given in Fig. 3.1).

References

Findlay JBC, Geisow MJ (eds.) (1989) Protein sequencing: a practical approach. IRL Press, Oxford, p 123

Choli-Papadopoulou T, Skendros Y, Katsani K (1997) In: Kamp RM, Choli-Papadopoulou T, Wittmann-Liebold B (eds.) Protein structure analysis. Springer, Berlin, p 137

3.1.2 Thin-Layer Chromatography of Nucleoside Phosphates

The following solvents are suitable for chromatography of adenosine triphosphate (ATP) and derivatives:

Solvents

0.2 M ammonium hydrogencarbonate in H_2O
or 0.4 M formic acid in H_2O
or 0.2 M sodium hydrogenphosphate in H_2O

Use a sheet of polyethyleneimine cellulose (PEI Cellulose), 50 × 200 mm, for chromatography. Mark the start about 15 mm away of the shorter side with a pencil. Apply the sample with a pipet or syringe. If larger volumes have to be used, apply them in several portions and/or a short line.

Develop the chromatogram ascending in a chamber lined with filter paper and equilibrated with one of the solvents.

After drying the sheet with a cold fan ^{32}P-labeled nucleotides are identified by autoradiography or a PhosphoImager. Unlabeled substances are identified by fluorescence quenching (see below).

The most effective solvent is sodium hydrogenphosphate, if the detection is not done with phosphate reagent (e.g., Hanes' reagent; see below).

3.1.3 Gradient Thin-Layer Chromatography of Nucleotides

Solvents

1 0.7 M ammonium formiate in H_2O
1′ 2 M ammonium formiate in H_2O
1″ 3 M ammonium formiate in H_2O
2 0.2 M lithium chloride in H_2O
2′ 1 M lithium chloride in H_2O
2″ 1.6 M lithium chloride in H_2O

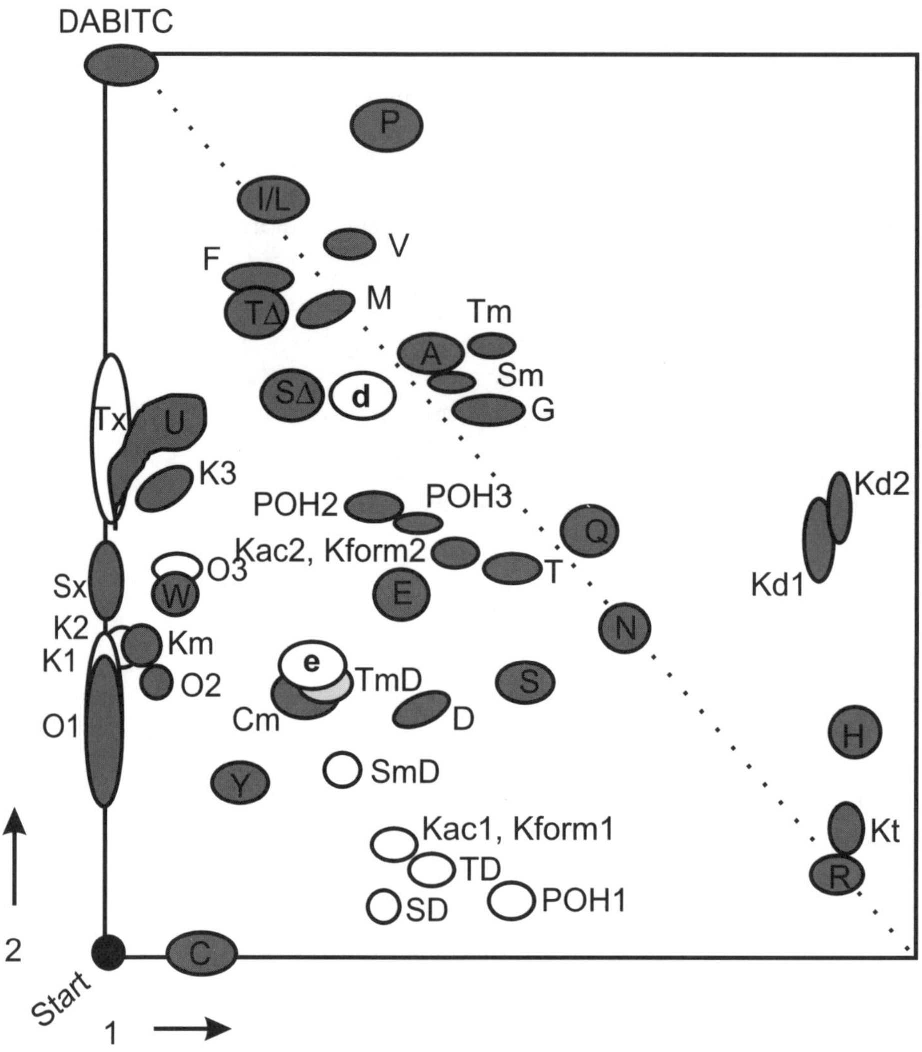

Fig. 3.1. Schematic drawing of 2D-TLC of DABTH derivatives of amino acids (modified from CHOLI-PAPADOPOULOU T et al. (1997). In: KAMP RM et al. (eds.) Protein structure analysis, p. 146. Springer, Berlin). *A* alanine, *C* cysteine, *Cm* caboxymethyl cysteine, *D* aspartic acid, *E* glutamic acid, *F* phenylalanine, *G* glycine, *H* histidine, *I* isoleucine, *K1* α-DABTH-ε-DABTC-lysine, *K2* α-PTH-ε-DABTC-lysine, *K3* α-DABTH-ε-PTC-lysine, *Kac1* α-DABTZ-N-ε-acetyl-lysine (after hydrolysis of acetyl group), *Kac2* α-DABTH-N-ε-acetyl-lysine, *Kform1* α-DABTZ-N-ε-formyllysine (after hydrolysis of formyl group), *Km* N-ε-methyl-lysine, *Kd* N-ε-dimethyllysine, *Kt* N-ε-trimethyllysine, *L* leucine, *M* methionine, *N* asparagine, *O1* α-DABTH-ε-DABTC-ornithine, *O2* α-PTH-ε-DABTC-ornithine, *O3* α-DABTH-ε-PTC-ornithine, *POH1* DABTZ-4-hydroxyproline, *POH2* 4-hydroxyproline, *POH3* 4-hydroxyproline (after hydrolysis), *Q* glutamine, *R* arginine, *S* serine, *SΔ* dehydroserine, *Sx* polymerized serine, *SD* DABTZ-serine, *Sm* O-methyl-serine, *SmD* DABTZ-O-methyl-serine, *T* threonine, *TΔ* dehydrothreonine, *TO* DABTZ-threonine, *Tx* polymerized threonine, *Tm* O-methyl-threonine, *TmD* DABTZ-O-methyl-threonine, *U* thiourea derivative, *V* valine, *W* tryptophan, *Y* tyrosine, *d* and *e*, reference marker of DABITC-reaction products of diethylamine and ethanolamine, respectively. *DABITC* 4-N,N′ dimethylaminoazobenzene-4′-isothiocyanate; *DABTH* diaminoazobenzene-thiohydantoin, *DABTZ* dimethylaminoazobenzene-thiazolinone, *PITC* phenylisothiocyanate, *PTH* phenylthiohydantoin

Formiate Gradient

Mark the start with a pencil on a PEI Cellulose sheet (dimensions as in Protocol 3.1.2) and apply the samples with a microliter syringe or a pipet.

Develop the sheet with H_2O for about 1.5 h to a height of 16 cm, then dry it and continue development with Solvent 1 to a running distance of 13 cm. Move the sheet in methanol for some minutes and dry it again. Then do the same procedure with Solvent 1′, followed by a third run with Solvent 1″.

Detect the phosphates as outlined in Protocol 3.1.2.

Lithium Chloride Gradient

If lithium chloride gradient is used, develop the sheet ascending for 2 min with Solvent 2, and then for 6 min with Solvent 2′ and last up to 16 cm with Solvent 2″.

Depending on the nucleosides, which of the solvent systems gives the best results should be verified.

3.1.4 Identification of Phosphates on TLC Plates

Solutions/Reagents

A 0.01% fluorescein sodium (w/v) in ethanol

B HANES' reagent: 0.5 g ammonium molybdate are dissolved in 5 ml H_2O and supplemented with 2.5 ml of 25% HCl followed by 2.5 ml of 70% perchloric acid. After cooling acetone is filled up to 50 ml. The reagent is ready for use the next day and it is stable for 2–3 weeks in the dark.

Identification by Phosphorescence

Phosphorescence is observed when the dry sheet is dipped in a container with liquid nitrogen and illuminated by a UV lamp. Purine derivatives phosphoresce light blue; pyrimidines are not visible.

Identification by Fluorescence Quenching

To detect the spots on the TLC plate spray it with Soln. A and dry it. Illumination with a UV lamp (254 nm) shows the spots dark in a greenish fluorescent surrounding. Indicate the spots with a pencil.

Identification by Molybdatophospho Complexes

Spray the dry plate with Soln. B and dry with air and heat in a drying oven to 110 °C until blue spots of phosphoric acid esters, e.g., phospholipids, appear.

References

Touchstone JC (1992) Practice of thin layer chromatography. 3rd. ed. Wiley, New York

3.1.5 Lipid Extraction and TLC of Lipids

Solutions/Reagents

A chloroform/methanol 1:1 (v/v)
B chloroform/methanol 3:1 (v/v)
C chloroform/methanol/1,2 N HCl 10:10:1 (v/v/v)
D 3% copper acetate (w/v), 8% phosphoric acid (w/v) in ddH_2O

Mobile solvents

1 chloroform/methanol/4.3 M ammonia 90:65:20 (v/v/v)
2 n-propanol/4.3 M ammonia 65:35 (v/v)
3 n-butanol/chloroform/glacial acetic acid/ddH_2O 60:10:20:10 (v/v/v/v)
4 isopropylether/glacial acetic acid 96:4 (v/v)
5 petrol ether/diethylether/glacial acetic acid 90:10:1(v/v/v)
6 chloroform/methanol/ddH_2O 65:25:4 (v/v/v)
7 n-butanol/glacial acetic acid/ddH_2O 60:20:20 (v/v/v)
8 chloroform/methanol/acetone/glacial acetic acid/ddH_2O 75:15:30:15:7.5 (v/v/v/v/v)

Extraction

Freeze the tissue, which should be extracted, with liquid nitrogen, and pulverize it in a mortar under liquid nitrogen.

Extract the tissue powder with 1 ml/g of Soln. A in a glass/Teflon homogenizator at RT.

Centrifuge at room temperature, decant the liquid, extract with 10 vol. of Soln. B, and centrifuge again and homogenize with 2.5 vol. of Soln. C again.

Combine the liquid extracts and vaporize the solvent in a nitrogen atmosphere.

Thin-Layer Chromatography

Activate the silica gel TLC plates for 3 h at 120 °C and store them in a desiccator.

Apply the lipid-containing solution as a short line with a microliter syringe or a pipet. The start zone should be as thin as possible. If larger volumes have to be applied, pipet several times onto the same position with intermediate drying. For drying the starting area as well as the plate after chromatography use a stream of nitrogen instead of air to avoid oxidation of unsaturated lipids.

Develop the chromatogram one- or two-dimensionally using the solvents 1–8. The choice of solvents depends on the lipids to be analyzed and has to be checked experimentally.

Phosphatidylinositols are separated on a gypsum-free silica plate, saturated with a 1% potassium oxalate solution in water, and activated as described above. The recommended solvent is Soln. 2. A second run with the same solvent after an intermediate drying increases the separation performance.

A general chromatographic system for separation of lipids does not exist. Whereas sophisticated systems for groups of lipids are described in literature, do not hesitate to try other solvent systems, multiple developments or a 2D chromatography to solve your distinct separation problem.

Identify the unknown lipids by comparison with a set of known lipids (standards) of the same type, chromatographed in the same run on the same plate.

Lipid detection

A versatile detection reagent for lipids is iodine vapor. The dry plate is placed into a chamber containing one or two Petri dishes with crystals of solid iodine. Lipid-containing spots are colored yellow to brownish after few minutes. The color disappears quickly in contact with air.

An additional general detection method is the oxidation by concentrated sulfuric acid. Spray the dry plate in a hood with concentrated sulfuric acid and then heat it in a drying oven to 130 °C. Organic substances yield dark brown spots.

Lipids are visualized by spraying the plate with Soln. D and heating to 180 °C 15 min.

For identification of phospholipids use HANES' reagent as described in Protocol 3.1.4.

For quantification of phospholipids identify the spots with iodine vapor or Soln. D. Mark the spots with a pencil, and cut the spots as well as a part of the plate with the same size but free of lipids. Transfer the material into a test tube and determine the amount of phosphate as described in Protocol 1.3.3.

References

Lowenstein JM (ed.) (1969) Lipids. Meth Enzymol 14

Neuhoff V et al. (ed.) (1973) Micromethods in molecular biology. Springer, Berlin

Touchstone JC, Dobbins MF (1983) Practice of thin layer chromatography, 2nd ed., Wiley, New York

3.2 Hints for Column Chromatography of Proteins

Column chromatography is used to analyze and prepare biomolecules. Chromatography techniques separate them according to differences in their specific properties as size, charge, hydrophobicity, and biospecific interactions. All chromatographic techniques use these properties and are different only with respect to instrumentation. The following hints are made for chromatography on columns of packed beds (matrix, stationary phase, support) passed by solvents (mobile phase) driven by low or medium pressure. The main difference to high-performance liquid chromatography (HPLC) is given by particle size and hydrodynamic resistance resulting in much higher backpressures in the latter case. Theoretical considerations are the same in low-pressure chromatography and HPLC, but because HPLC needs much more complex technical equipment, this branch of chromatography requires more discussion, which is beyond the scope of this chapter. As a sole example, the analytical separation of carbohydrates by high-performance ion exchange chromatography (HPIEC) is described at the end of this chapter.

The success of chromatographic separations depends on skillful realization as well as proper selection of media, conditions, and separation principles.

Biologic macromolecules are molecules bearing a lot of different properties. Mostly these properties are not distributed uniformly on the surface of the molecule. Each of these parts of the molecules is able to interact with chromatographic media; so the stationary and mobile phases produce a special interaction, but other properties may also contribute to separation. Macroscopic properties of a macromolecule as isoelectric point, amino acid composition, posttranslational modifications, such as, for example, glycosylation, acylation by fatty acids, or phosphorylation, may favor a type of chromatography, but mostly other kinds of interaction have to be suppressed during a distinct chromatographic process. Therefore, it is impossible to give a single protocol for a fast and selective separation with high yield of the wanted macromolecule.[1]

But some general rules are possible:

If several chromatographic steps are planned, the same type of chromatography should never follow the preceding, i.e., when the first step was gel chromatography then ion exchange or affinity chromatography should follow, or gel filtration should be used after ion exchange. The reasons are, for example, the concentration of the sample in ion exchange after dilution by gel filtration and the ionic strength caused by ion exchange is reduced by gel filtration, respectively. So additional buffer exchanges can be avoided.

For selection of a chromatographic method it should be taken into consideration whether the protein of interest may be denatured (or if it can be renaturated) or some specific properties as ligand binding or enzymatic activity must be conserved during purification. These reflections are not relevant, if during analytical separation a signal produced by a covalently attached label is measurable independent of the structure of the macromolecule.

The following checklist helps in planning a purification procedure:

- What is the source for preparation of the molecule of interest?
- Which kind of tissue maceration is optimal with respect to conservation of properties and yield?
- Is it possible to start with non-chromatographic procedures, e.g., centrifugation?

[1] Illustrative handbooks describing the different types of low-pressure chromatography are found as .pdf files, e.g., at www.chromatography.amershambiosciences.com.

- Is the macromolecule of interest incorporated into organelles, e.g., inclusion bodies, membranes, which afford (partial) denaturation?
- Are properties of the molecule of interest inhibited by components of the buffers (e.g., inhibition of hem proteins by sodium azide or enzyme activity by SDS)?
- Is the monitoring of the purification process disturbed by buffer components?
- Are properties of the molecule of interest regulated by divalent cations and do chelators as EDTA therefore affect them?
- Is it necessary to add a stabilizator, such as, for example, sucrose or glycerol?

The following of protocols for purification of similar macromolecules is very often successful; nevertheless, it is necessary to optimize the procedure because it is possible that the given protocol will miss some essentials, or that the wanted macromolecule will differ in important properties.

Before starting the purification of a more or less known macromolecule, it is helpful to collect information by the following:

Analytical methods

- Determination of molar mass by SDS-PAGE
- Determination of isoelectric point
- Identification of posttranslational modifications
- Determination of enzymatic parameters such as substrate specificity, inhibitors, activators, pH optimum

Investigation of buffer compatibility

- Inhibition by buffer components
- Selection of stabilizators
- Renaturation after application of urea, guanidinium hydrochloride, or alcohols
- Stability in the presence of detergents

Test of possible chromatographic procedures

- Binding to specific ligands
- Binding to cation or anion exchanger at distinct pH and ionic strength or to hydrophobic supports (binding may be as useful as no binding if impurities show the opposite behavior)
- Significant differences in molecular size if gel filtration is intended

As an example for combination of different methods in protein purification Fig. 3.2 gives the flow chart for the isolation of membrane protein complex.

The classification of chromatography as gel permeation chromatography (GPC; size-exclusion chromatography, SEC; gel filtration, GF), ion exchange chromatography (IEC), hydrophobic interaction chromatography (HIC), and affinity chromatography (AC) is

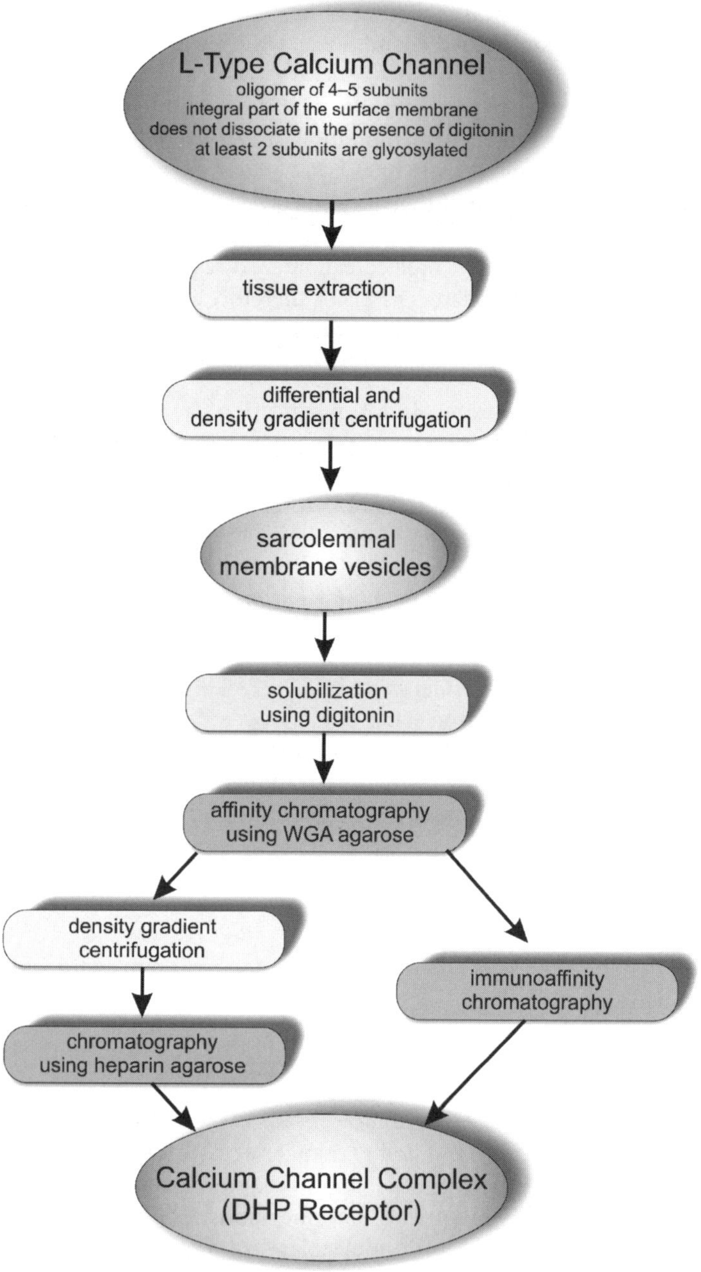

Fig. 3.2. Flow chart for the purification of the L-type calcium channel complex from porcine heart. (Modified from HAASE et al. 1991, Eur J Pharmacol 207:51)

based on theoretical models. Besides the main interactions between the stationary matrix and the biomolecule, which are the reason for classification, a couple of interactions exist as outlined above.

So you can observe ion exchange and hydrophobic interactions in GPC, and size-exclusion effects are not negligible if macroporous ion exchange supports are used.

Substances destroying the water structure, the so-called chaotropic substances such as urea, guanidinium hydrochlorid, or some organic solvents such as methanol, ethanol, or acetonitrile, suppress hydrophobic interactions. Ions also alter the water structure. The power of destruction of water structure is given by the Hofmeister series:

$$\leftarrow \quad \text{increasing precipitation ("salting-out") effect}$$

$$NH_4^+, Rb^+, K^+, Na^+, Cs^+ > Li^+ > Mg^{++} > Ca^{++} > Ba^{++}$$

$$PO_4^- > SO_4^- > CH_3COO^- > Cl^- > Br^- > NO_3^- > ClO_4^- >$$

$$> F_3CCOO^- > I^- > SCN^-$$

$$\text{increasing chaotropic ("salting-in") effect} \quad \rightarrow$$

Besides the application in chromatography, chaotropic substances are used for protein extraction, e.g., of recombinant proteins from inclusion bodies and for precipitation by salts or organic solvents (cf. Protocols 3.7.2 and 3.7.3).

References

Deutscher MP (ed.) (1990) Guide to protein purification. Meth Enzymol 182

Gooding KM, Regnier FE (eds.) (1990) HPLC of biological macromolecules: methods and applications. Chromatogr Sci Series, vol. 51. Marcel Dekker, New York

Harris EL, Angal S (1990) Protein purification methods: a practical approach. Oxford Univ. Press, Oxford

Holtzhauer M (ed.) (1996) Methoden in der Proteinanalytik. Springer, Berlin, p 7

Mikes O (1988) High-performance liquid chromatography of biopolymers and biooligomers. Part A: principles, materials and techniques; Part B: Separation of individual compound classes. J Chromat Library. Elsevier, Amsterdam, vol. 41A and vol. 41B

Oliver RWA (ed.) (1989) HPLC of macromolecules: a practical approach. IRL Press, Oxford

3.3 Gel Permeation Chromatography (GPC; Gel Filtration, GF; Size-Exclusion Chromatography, SEC)

Gel permeation chromatography is the mildest of all chromatographic techniques. It allows to select the buffer conditions during separation so that no alteration of the protein structure or function occurs.

The foundation of this technique is the penetration of molecules into the cavities of a macroporous support particle. These particles are mostly made from hydrophilic gels of dextran, agarose, polyacrylamide, or chemical derivatives of them, and the interactions between the macromolecules of the chromatographic support and the biomolecules are minimized. At given diameters of the cavities of the chromatographic matrix, molecules with a hydrodynamic diameter similar to or smaller than the pore diameter diffuse into the matrix, whereas molecules with larger diameters are excluded. Because in a rough approximation the diameter of a protein is proportional to its molar mass, gel filtration is used for separation by molecular weight (in a more detailed view, separation according to molar mass is only possible if all molecules have the same shape, e.g., spherical or ellipsoid).

GPC can be applied in two distinct ways:

Group separations: One group of molecules of a sample is excluded from the gel, whereas the other group penetrates the pores of the matrix. Examples: desalting; buffer exchange.

Fractionation according to molecular size ("molar mass"): The components of a sample are separated according to their molecular size caused be their ability to penetrate the macoporous beads. Examples: Analytical or preparative fractionation of macromolecules, separation of oligomers from monomers.

Most of the supports are chemically stable at pH between 2 and 12 and are not influenced by salt. Especially the gels for low-pressure chromatography are sensitive to organic solvents at higher concentrations. Especially gels with high exclusion limit also collapse irreversibly at higher pressure. Supports for medium and high-pressure chromatography show the same chemical stability, but they have much smaller particle diameters and are stable at pressures of some hundred bar. The mechanical instability of the soft swollen gels forbids the use of magnetic stirrers or fast moving blade agitators.

Swelling of dry GPC beads takes up to several hours and may be accelerated by heating (cf. Table 3.2). Take care that sufficient solvent is available during swelling (as a rule of thumb twice the expected volume).

To avoid irreversible compression of the gel, respect manufacturer's specifications of maximal pressure during filling of the column and during run. (For selected gels, this value is given in Table 3.2.) The hydrostatic pressure Δh during filling and run is illustrated in Fig. 3.3.

If pumps, e.g., peristaltic pumps, are used for the chromatographic run, occurring changes in viscosity result in an unexpected increase of pressure; therefore, select flow rates significantly lower than the maximal rate given for a distinct gel.

To compare the separation performance of two columns containing the same media, the flow rate of the solvent (mobile phase) is an important quantity. Because it is easy to measure, the flow rate is often given as volumetric flow rate V with the dimension milliters per minute. But this quantity depends on the diameter of the column; therefore, it would be better to give the linear flow rate L in centimeters per hour, which eliminates the cross-section of the columns:

$$L = \frac{V \cdot 240}{\pi \cdot d^2}$$

L: linear flow rate ($cm \cdot h^{-1}$); V: volumetric flow rate ($ml \cdot min^{-1}$); d: inner diameter of the column (cm).

When mechanically stable gels are used, the linear flow rate L is proportional to the applied pressure (DARCY's law):

$$L = \frac{C \cdot \Delta h}{H}$$

L: linear flow rate in centimeters per hour; C: constant of proportionality dependent on the properties of the bed material and the applied buffer; Δh: hydrostatic pressure in centimeters (cf. Fig. 3.3); H: bed height in centimeters.

For Sephadex the following values of proportionality constant C are given in Table 3.1.

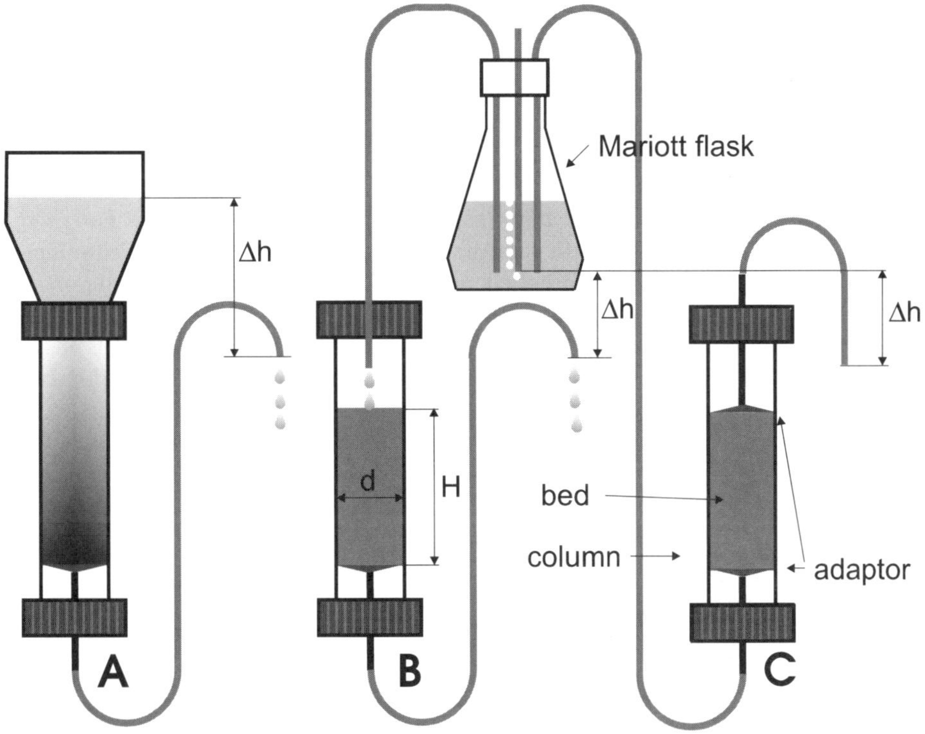

Fig. 3.3. Column chromatography. Principle of column filling and variants of chromatographic run (for descriptions see text)

Table 3.1. Constants of proportionality for Sephadex GPC media

Gel type	C	Particle diameter (μm)
G-10	19	55–165
G-25 superfine	9	17–70
G-25 fine	30	35–140
G-25 medium	80	85–260
G-25 coarse	290	170–520
G-50 fine	36	40–160

Data from Amersham Biosciences Handbook, Gel filtration: Principles and methods 18-1022-18. Edition AI

The best column design for GPC as well as for affinity chromatography is a column with adaptors at both ends (Fig. 3.3C). This column type guarantees even surfaces of the gel bed, symmetrical application of sample, and regular flow of the eluent.

The quality of separation depends on several factors. One point is the size and size distribution of the matrix beads. Small beads of same size increase fractionation efficacy, but also increase the flow resistance; therefore, fractionations should be made with beads characterized as "medium," "fine," or "superfine," whereas group separations are mostly successful with "coarse" media.

A second point of view is the geometry of the column. For fractionations the ratio of bed height to bed diameter should be at least 50:1, but this value is a rule of thumb and may be varied depending on the geometry of adaptors and size distribution of matrix beads.

The third point concerns the size of the sample. For group separations the sample volume may be up to 1/3 of the bed volume, whereas for fractionations the sample volume should be on the order of 1/50 of the bed volume.

A fourth factor is the flow rate of the eluent (mobile phase). A too high rate decreases resolution because there is no time for molecules to diffuse into the pores of the matrix. In contrast, a very slow migration of solvent decreases the resolution by remixing the components by diffusion. The effect of diffusion is minimized if the chromatography is done at low temperature. If short separation times are necessary, pre-packed columns, elevated pressure, or HPLC columns are indispensable.

3.3.1 Selection of Supports

The main criterion for selection of a GPC matrix is the fractionation range. Manufacturers of GPC media declare a fractionation range and avoid molar weight. These values are mostly obtained by chromatography of defined homopolymers such as polysaccharides or polystyrenes, or with a special set of protein standards. Since a gel filtration matrix separates according to steric (hydrodynamic) parameters, and an ellipsoid molecule has a larger hydrodynamic radius than a spherical one with the same molar mass, both the molecules therefore show different retention behavior. If the protein of interest is significantly larger than accompanying substances, a matrix with a cutoff between them is recommended because the target appears with the void volume as the first little diluted fraction. So, for desalting of protein solutions or for buffer exchange, mostly Sephadex G-25 medium is well suited.

Additional chemical and mechanical stability of supports should be taken into consideration, as well as possible interactions between protein and matrix macromolecules.

3.3.2 Filling of a Gel Filtration Column

The GPC column, which is intended for fractionation in laboratory scale, should have a ratio of bed diameter d to bed height H (Fig. 3.3B) of 1:50 to 1:200.

Slurry the swollen, settled gel in about a half of its volume of buffer and degas under vacuum for 10 min. Connect the column outlet with tubing ending in a vessel at the level of the upper border of the column. Then mount the filling reservoir to the vertical column as shown in Fig. 3.3A. Fill the column up to one-fourth to one-third of its height with buffer and pour the gel slurry without bubbles in one portion into column and reservoir. Regulate the recommended hydrostatic pressure Δh (cf. Fig. 3.3; Table 3.2) by lowering the outlet.

When the gel has settled, fill the column with buffer and carefully insert the upper adaptor. Move the adaptor down to the surface of the packed bed and push it further 1–3 mm into the GPC gel. Avoid air bubbles and let the outlets open to avoid compression of the gel.

If columns without adaptors are used, stabilize the surface of the gel by a sheet of (glass fiber) filter paper. The surface of the bed should be without craters.

The GPC columns with bubbles or clefts within the bed have to be emptied and packed again prior use.

Wash the completely packed column with buffer (eluent) until a constant flow rate is obtained. The eluent volume should be at least fivefold that of the bed volume.

3.3.3 Sample Application and Chromatographic Separation (Elution)

For fractionation the sample volume should be not more than 1/20 to 1/10 of bed volume and should have the shape of a cylinder.

The smaller the height of the sample, the better is the resolution. But a compromise has to be made between concentration and volume, because highly concentrated protein solutions tend to form precipitates on gels.

The sample may be applied by several ways:

- If adaptor columns are used, the eluent flow is interrupted and the sample is applied via a T-piece. Be sure that no air is introduced.
- Some buffer overlies the bed and the sample, which has a higher density made by supplementing with salt or sucrose; it is added to form a lower layer.
- Stop the eluent flow when the meniscus of the supernatant liquid has reached the bed surface. Carefully add the sample and avoid whirling up the gel, open the outlet until the liquid has retracted and then start the elution.

After applying the sample, connect the column with the elution buffer. If columns without adaptor are used, leave some millimeters of buffer above the bed and end the solvent tubing directly at the surface of the buffer zone.

A constant hydrostatic pressure during chromatography is achieved if a Mariotte flask (Fig. 3.3) is used. Regulate the flow rate by varying Δh and/or by a pinch clamp at the outlet.

To avoid re-mixing of separated components, make the distance between column outlet, detector, and fraction collector as short as possible. For normal pressure chromatography use polypropylene or Teflon tubing with an inner diameter of about 1 mm.

If adaptor columns and pumps are used, it is unimportant whether the eluent flows from top to bottom of the column, or vice versa. In the case of very soft gels with high exclusion limit, choose a run against gravity from bottom to top.

Instead of a sole very long column, which is very difficult to fill, it is possible to take several shorter columns connected inlet to outlet. For this construction adaptor columns are necessary and the middle column runs from bottom to top.

3.3.4 Cleaning and Storage

GPC supports based on dextran (e.g., Sephadex) are cleaned from residual proteins by rinsing with five to ten bed volumes of 2 M NaCl, 0.1 M NaOH, or detergent as 0.1% Tween 20. Equilibrate the column after cleaning with ten bed volumes of elution buffer. Agarose-based beds, e.g., Sepharose or Biogel A, are freed from pollutions by sodium chloride, detergent, or 6 M urea.

Store the GPC media at neutral pH in a refrigerator in the presence of an antibiotic, e.g., 0.02% sodium azide or 0.1% Thimerosal, or in 20% ethanol or i-propanol.

3.3.5 Determination of Void Volume V_0 and Total Volume V_t

Solutions/Reagents

A PBS
B 2 mg/ml dextran blue (mean $M_r > 2000$ kD) in PBS
C 1 mg/ml dinitrophenylalanine in PBS

Equilibrate a packed GPC column, e.g., Superdex 200, 0.9 cm diameter, 50 cm bed height, 32 ml bed volume, with 350 ml PBS. Adjust the flow rate to 30 cm/h (0.3 ml/min).

Apply 0.5 ml of Soln. B as described in part 3.3.3 and continue the elution with Soln. A. Monitor the elution either continuously by UV measuring at 280 nm or by collecting 1-ml fractions, which are measured at 280 nm. Plot the optical density against elution volume (cf. Fig. 3.4). The void volume V_0 is the volume at maximal optical density.

For determination of the total volume V_t use Soln. C instead of Soln. B.

Calibrate a GPC column for molar weight determination analogously. Use a set of defined proteins or polypeptides as standards (see inset in Fig. 3.4). If the mass differences allow a complete separation, the proteins can be applied as a mixture.

For characterization of a GPC medium other than defined macromolecules as dextrans, polypeptides, polystyrene sulfates, or colloidal gold may be used. For these standards the eluent is ddH_2O. These macromolecules are suitable for molar mass determination of proteins with restrictions only.

References

Amersham Biosciences (2002) Gel filtration: principles and methods (www.chromatography.amershambiosiences.com)
Stellwagen E (1990) In: Deutscher MP (ed.) Guide to protein purification. Academic Press, San Diego, p 317

3.3.6 Removing of Unbound Biotin After Conjugation by Gel Filtration ("Desalting")

Instead of dialysis, small molecules, such as salts or organochemical substances, can be separated quickly from macromolecules by gel filtration. Buffer exchange is also possible this way. The disadvantage is a dilution of the sample.

The following example illustrates the fast removal of unbound biotin after conjugation to a protein (Table 3.2).

Solutions/Reagents

A 2 mg/ml dextran blue (mean $M_r > 2000$ kD), 0.2 mg/ml Ponceau S in PBS
B 0.02% sodium azide (w/v) or 1% Neolone (v/v) in PBS
PBS

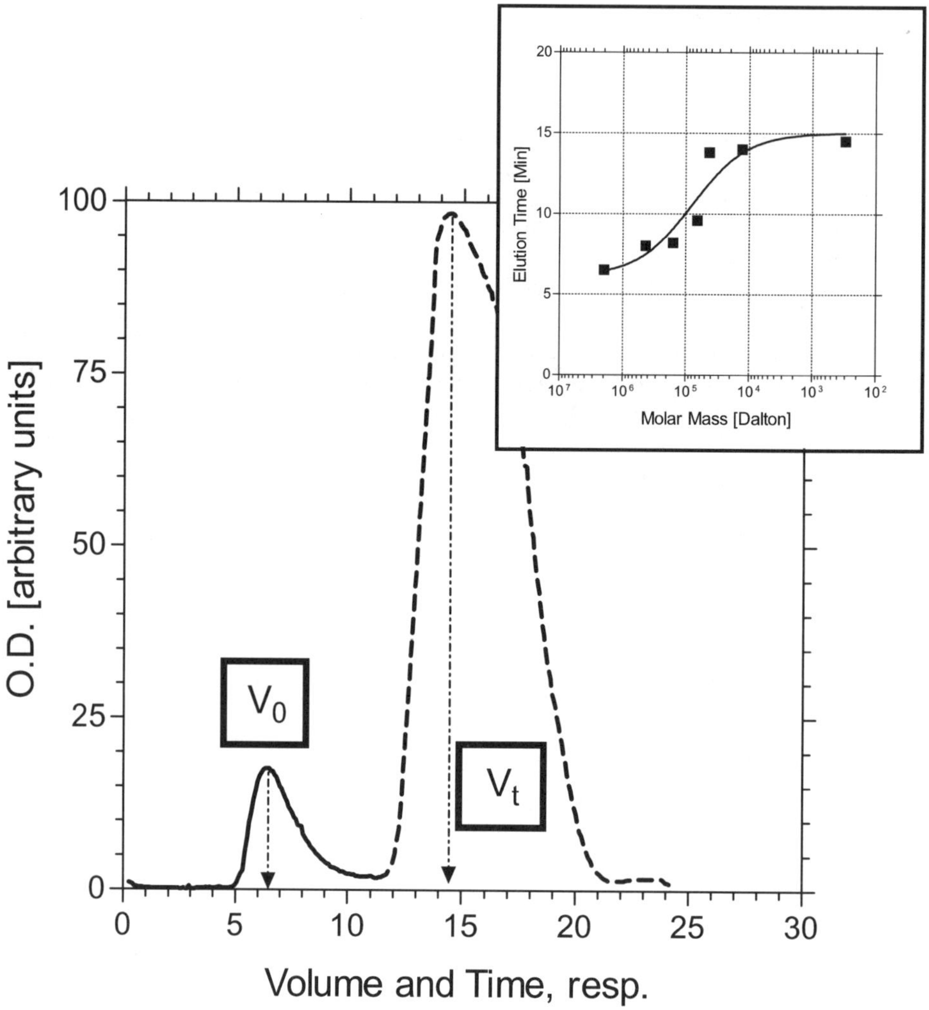

Fig. 3.4. Determination of V_0 and V_t by gel permeation chromatography. *Dashed line:* dextran blue (defines void volume V_0); *dotted line:* DNP-alanine (defines total volume V_t). *Inset:* plot of a calibration curve for determination of molar mass

Swell 0.5 g Sephadex G-25 in 5 ml PBS for 3 h at RT. Then fill 1 ml of the settled gel into an empty column (0.5- to 0.8-cm diameter) and wash the bed with 20 ml PBS. Remove excess buffer from the bed surface after washing.

Determine V_0 and V_t by chromatography of 0.5 ml of Soln. A. The blue fractions represent that volume which will contain the biotinylated conjugate, and the red fractions will give the waste (unbound biotin). After chromatography, equilibrate the column again with PBS.

Dilute the mixture of biotinylation to 0.5 ml with PBS. Carefully pour the solution onto the surface of the bed, allow the mixture to

Table 3.2. Selected media for GPC and IEC

Gel type		Bed volume (ml/g)	Fractionation range (kD)	Remarks
GPC media				
Sephadex[a]	G-10	2–3	To 0.7	Swelling time 3 h (20°)
	G-25	3–4	1–5	Swelling time 3 h (20°)
	G-50	9–11	1.5–30	Swelling time 3 h (20°)
	LH-20	1.4–4.6[b]	To 5	GPC in organic solvents
Superdex[a]	Peptide	Pre-swollen	0.1–7	
	30	Pre-swollen	To 10	
	75	Pre-swollen	3–70	
	200	Pre-swollen	10–600	
Superose[a]	6 HR	Pre-swollen	5–5000	
	12 HR	Pre-swollen	1–3000	
Sephacryl[a]	S-100 HR	Pre-swollen	1–100	
	S-200 HR	Pre-swollen	5–250	
	S-300 HR	Pre-swollen	10–1500	
	S-400 HR	Pre-swollen	20–8000	
Bio-Gel[c]	P-2	3	0.1–1.8	
	P-4	4	0.8–4	
	P-6	6.5	1–6	Exclusion limit ds-DNA 6 bp
	P-10	7.5	1.5–20	
	P-30	9	2.5–40	Exclusion limit ds-DNA 22 bp
	P-60	11	3–60	Exclusion limit ds-DNA 55 bp
	P-100	12	5–100	
	A-0.5 m Gel	Pre-swollen	10–500	
	A-1.5 m Gel	Pre-swollen	10–1500	
	A-5 m Gel	Pre-swollen	10–5000	
	A-15 m Gel	Pre-swollen	40–15 000	
	A-50 m Gel	Pre-swollen	100–50 000	Exclusion limit ds-DNA 350 bp
Macro-Prep[c]	SE 100–40	Pre-swollen	5–100	
	SE 1000–40	Pre-swollen	10–1000	
BioSil[c]	SEC 125–5	Pre-swollen	5–100	
	SEC 250–5	Pre-swollen	10–300	
	SEC 400–5	Pre-swollen	20–1000	
Ion exchange media			Capacity (mg/ml)[d]	
Sephadex[a]	DEAE A-25		30 (HSA)	Anion exchanger
	DEAE A-50		110 (HSA)	Anion exchanger
	QAE A-25		10 (HSA)	Anion exchanger
	QAE A-50		80 (HSA)	A nion exchanger

Table 3.2. (continued)

Gel type		Bed volume (ml/g)	Fractionation range (kD)	Remarks
	CM C-25		1.6 (IgG)	Cation exchanger
	CM C-50		7.0 (IgG)	Cation exchanger
	SP C-25		1.1 (IgG)	Cation exchanger
	SP C-50		8.0 (IgG)	Cation exchanger
Sepharose[a]	Q		70 (BSA)	Anion exchanger
	SP		55 (RNase)	Anion exchanger
	DEAE		110 (HSA)	Anion exchanger
	CM		15 (IgG)	Cation exchanger
	DEAE CL-6B		170 (HSA)	Anion exchanger
	CM CL-6B		9.5 (IgG)	Cation exchanger
Sephacel[a]	DEAE		160 (HSA)	Anion exchanger
Mono Q[a]			65 (HSA)	Anion exchanger
Mono S[a]			75 (IgG)	Cation exchanger
Bio-Scale[c]	Q		20 (BSA)	Anion exchanger
	DEAE		25 (BSA)	Anion exchanger
	S		50 (IgG)	Cation exchanger

[a] Supplier: Amersham Biosciences
[b] Depending on used organic solvent
[c] Supplier: BioRad
[d] Dynamic capacity, dependent on bound protein

penetrate into the bed, and elute two times with 0.3 ml PBS each. The combined eluates contain the biotinylated conjugate.

Wash the column with 20 ml of Soln. B and store it in a refrigerator.

3.4 Ion Exchange Chromatography (IEC)

In ion exchange chromatography, electric charges are immobilized to insoluble supports. If positive charges, realized by alkylated amino residues, are immobilized, the chromatographic medium acts as an anion exchanger, and immobilized anions, such as carboxyl or sulfonic acid groups, result in cation exchange media.

Ion exchange chromatography, hydrophobic interaction chromatography (HIC), and affinity chromatography (AC) show some similarities concerning practical realization; therefore, most of the hints given for IEC are applicable for HIC and AC. Examples for AC are given in Protocols 3.6.2.4 (biospecific desorption) and 3.6.2.5 (elution by partial denaturation).

A problem of IEC as well as of AC is a decline of capacity during repeated use of the same column. Especially anion exchangers containing quaternary amines (QAE media) suffer from chemical decomposition of the charged groups, whereas in AC proteinous ligands tend to denaturation (loss of biospecifity) during reuse.

3.4.1 Preparation of Ion Exchange Supports

Independent from the chemical nature of the supports, pre-cycle IEC media before use to obtain optimal separations. The pre-cycling is recommended especially before first use and after longer storage.

Swell the IEC medium in first solution (see Table 3.3). For dry media use 15 ml liquid per gram support, in the case of pre-swollen media a ratio of 5 ml per gram wet weight is sufficient. Equilibrate with gentle agitation for 30–90 min at RT (do not use magnetic stirrers!). Then suck the gel on a filtering funnel and wash with ddH_2O until the intermediate pH is reached (cf. Table 3.3). Slurry the gel in the second solution with a ratio of 1 g wet gel to 5 ml liquid. Suck the gel off after 30 min and wash with ddH_2O to nearly pH 7.

After pre-cycling equilibrate the IEC medium with starting buffer. Because a simple washing with a tenfold bed volume of starting buffer is not sufficient, use one of the both following methods.

Method A

Suspend the gel in a buffer with the same pH as the starting buffer, but tenfold concentrated with respect to the buffering counterion (anion exchange medium, e.g., phosphate). Pour the gel into the column after 15 min and wash it with 10 ml starting buffer per gram wet weight.

Method B

Suspend the gel in the starting buffer. Adjust the pH of the slurry after 10 min with acid or base until the pH of the starting buffer is obtained. When the pH is stable, pour the gel into a column

Table 3.3. Cyclization of ion exchange supports

IEC medium	First solution	Intermediate pH	Second solution
a) Cellulose or resinous IEC media			
Cation exchanger (e.g., CM, SP, SE)			
	0.5 N NaOH	≈ 9	0.5 N HCl
Anion exchanger (e.g., DEAE, PEI, QAE)			
	0.5 N HCl	≈ 5	0.5 N NaOH
b) Dextran- or agarose-based IEC media			
Cation and anion exchanger			
	0.1 N NaOH	≈ 9	1 M acetic acid

and wash until the conductance of the eluate has the value of the starting buffer (eluent).

3.4.2 Capacity Test

The capacity indicated in Table 3.2 is for orientation only and differs from protein to protein; therefore, check the capacity in an assay with the same conditions as for separation and monitoring.

The decision on use of an anion or cation exchange medium depends on binding of the protein of interest or binding of accompanying substances to the matrix. If the protein, which should be absorbed, has an isoelectric point (pI) below pH 7 (acidic protein), select an anion exchanger. If the absorbed molecule carries only few charged groups on its surface, a strong exchange medium is recommended (SP and QA). If the macromolecule carries a high charge density or if desorption (elution) with smooth conditions is favored, select a weak ion exchange medium (CM and DEAE).

For binding choose a buffer with low ionic strength and pH at least one unit below pI for cation exchange and one pH unit above pI for anion exchange, respectively.

If the pH dependency of biological stability and/or activity is known, a cation exchanger is recommended if this stability is below the pI.

Check the conditions for binding (adsorption) to the IEC gel as follows:

Equilibrate a series of 0.1 – 0.5 ml of ion exchange medium in test tubes with 50 mM buffer without any additional neutral salt. The buffers should differ in pH for 0.5 units each.

Add 0.5 – 1 ml of protein sample, dialyzed against the respective buffer, to the wet media. Incubate for 15 – 30 min at that temperature which should be used for separation. Separate by centrifugation or filtration after incubation and determine the amount of target protein in the liquid. For binding buffer select that buffer at which the target protein disappears.

The capacity of the exchange medium is determined with a series of protein dilution in binding buffer and a constant amount of IEC gel.

For preparative separations, take a gel volume such that its capacity is required for only two-thirds of maximal binding.

3.4.3 Sample Application

The binding of proteins to IEC media follows the law of mass action; therefore, complete binding is not possible; bound and free proteins are in equilibrium.

Loading of supports can be done in two ways: batch loading and column loading.

Batch loading

Mix the protein dissolved in binding buffer with equilibrated IEC gel. Agitate for several minutes and separate gel and solution either by centrifugation or filtration. Wash and elute the gel on a funnel or in a column. Since the equilibrium is established only once, a nearly complete absorption occurs if the gel is in a large excess. The advantage of batch loading is the easy separation of gel and liquid.

Column loading

Prepare a column with the equilibrated gel. In the opposite to GPC the ratio of diameter to bed height may be 1:1 and the sample volume may be much larger than the bed volume. A low bed height has the advantage of little change of height during elution when buffers with altered ionic strength or pH are used. Furthermore, the binding equilibrium establishes very often during movement of the sample through the column.

3.4.4 Elution

Gradient elution

Elution of bound material is achieved by changing the buffer composition, i.e., by increasing the ionic strength (addition of neutral salts as sodium or potassium chloride) or by changing the pH of the elution buffer. This change may be done by sequential addition of different elution buffers (stepwise gradient) or by application of a continous change (continous gradient) produced using a gradient mixer (cf. Fig. 2.2). In IEC mostly the basic buffer composition is retained and the ionic strength is increased by increasing the concentration of salt. The elution is monitored by UV measurement, and the development of the gradient is checked by measuring the conductivity and/or the pH.

It is possible to combine stepwise gradient and continuous gradient, i.e., the starting elution buffer has a higher ionic strength than the binding buffer, and after a short time the elution of the protein(s) of interest starts by application of a flat continuous gradient.

When soft IEC media are used, adaptor columns are not suitable because the volume of the gel alters (mostly shrinks) when salt concentration increases; hence, the flow direction is from top to bottom.

Flow rate of the elution buffer as well as change of buffer composition should not be too fast because the buffer components need some time to diffuse into the gel and proteins bound within the pores of the gel have to move outside.

3.4.5 Cleaning and Regeneration

After separation, wash the gel with two bed volumes of 2 M KCl or 6 M guanidinium chloride and re-equilibrate it. If this purification is not sufficient, cycle the gel as described in Sect. 3.4.1.

Store the gel in a refrigerator in the presence of a non-ionic biocide, e.g., alkylbenzoates (parabenes) or Neolone, or in 20% ethanol (v/v).

References

Rossomando EF (1990) In: Deutscher MP (ed.) Guide to protein purification (Meth Enzymol Vol. 182). Academic Press, San Diego, p 309

Amersham Biosciences (2004) Ion exchange chromatography: Principles and methods. ISBN 91 970490-3-4.
www.chromatography.amershambiosciences.com

3.4.6 High-Performance Ion Exchange Chromatography (HPIEC) of Mono- and Oligosaccharides

At strong alkaline conditions mono- and oligosaccharides are weak anions. So it is possible to separate them on alkaline-resistant HPLC columns. Whereas spectrophotometric monitoring is not possible, pulsed amperometric detection allows sensitive identification in the nanomolar range.

In the literature, isocratic solvent systems are found for monosaccharides, and gradient systems are described for oligosaccharides. Table 3.4 gives some gradient systems for carbohydrate separation; Table 3.5 illustrates programs for pulsed amperometric detectors.

The strong alkaline solvents need some special efforts concerning materials. Capillaries and fittings should be of PEEK. The injection valve must be designated to be alkali resistant (e.g., Tefzel rotors of Rheodyne valves). Storage bottles for mobile solvents have to be of plastics and should be supplemented with a tube filled with caustic soda pellets.

There is no need to degas solvents, but filtration using 0.45-µm Teflon filters is indispensable.

Table 3.4. Solvents for carbohydrate HPIEC

Flow rate (ml/min)	Eluent	Eluent program
Monosaccharides including amino sugars		
1	A: 16 mm NaOH; B: 200 mm NaOH	16 min 100% A, 5 min 100% B
Oligosaccharides		
1	A: 100 mm NaOH; B: 1 M Na acetate in A	5 min 2% B, 20 min 2% → 20% B, 10 min 30% B

Table 3.5. Program for pulsed amperometric detection (PAD)

	t_1	V_1	t_2	V_2	t_3	V_3	Integration time Start at	Integration time Duration
Monosaccharides[a]								
	1.2	0.05	0.2	0.7	0.5	−0.45	0.82	0.3
Oligosaccharides[b]								
	0.48	0.05	0.12	0.6	0.06	−0.6	0.15	0.3

Time, t, in seconds; potentials, V, in millivolts

[a] Biometra GmbH (1992) Instructions for pulsed electrochemical detector PAD 300

[b] Weizhandler M, Kadlecek DN, Avdalovic N et al. (1993) J Biol Chem 268:5121

References

Davies MJ, Hounsell EF (1998) HPLC and HPAEC of oligosaccharides and glycopeptides. In: Hounsell EF (ed.) Glycoanalysis protocols. 2nd ed. Humana Press, Totowa N.J., p 79
Delgado GA, Tenbarge FL, Friedman RB (1991) Carbohydr Res 215:179
Townsend RR, Hardy MR, Lee YC (1989) Meth Enzymol 179:65

3.5 Hydrophobic Interaction Chromatography (HIC)

At HIC (special case: reversed phase chromatography, rp chromatography), hydrophobic areas of polypeptides interact with immobilized alkyl or aryl residues of chromatographic supports. HIC sometimes is counted to affinity chromatography, but with respect to specifity and separation efficiency there are more similarities to IEC. Besides the binding mechanism, the main differences to IEC are binding and elution: In HIC binding takes place at elevated ionic strength and elution is done by lowering salt concentration or by addition of water-miscible solvents such as methanol, ethanol, propanol, or acetonitrile. These conditions are reasons for using HIC in desalting of protein solutions.

3.5.1 Capacity Test

Incubate 0.5 ml of the protein dissolved in a buffer with elevated ionic strength, e.g., 1 M ammonium sulfate or sodium chloride or potassium chloride, 50 mM phosphate buffer, and pH 7.0, with 0.5 ml of HIC media of different hydrophobicity, e.g., ethyl (C2), butyl (C4), hexyl (C6), and octyl (C8) substituted agarose, at 0 °C for 30 min. Centrifuge and analyse the supernatant for the protein of interest.

Depending on target stability or further steps of purification, especially with respect to the sometimes severe elution conditions, it has to be decided whether the interested protein will be adsorberd to the HIC matrix or stay in solution.

The HIC medium with the optimal properties is chosen for batch or column absorption.

Sample application is done analogously to ion exchange chromatography (see Sect. 3.4.3).

3.5.2 Elution

As described for IEC, elution is done by a stepwise or a continuous change of buffer composition. The mildest elution buffer is an aqueous buffer with low ionic strength, e.g., 20 mm Tris-HCl. If it is not successful, desorb with a chaotropic solvent, e.g., 2 M potassium rhodanide (thiocyanate), 2.5 M guanidinium hydrochloride, up to 7 M urea, or with increasing concentrations of methanol or acetonitrile. Especially the use of rhodanide or urea may be accompanied by a chemical modification of amino acid side chains, which disturbs amino acid analysis.

3.5.3 Regeneration

Regenerate the HIC medium after elution as follows:

Wash the gel with one bed volume of ddH_2O. Then wash with one bed volume each of 33, 67, and 95% ethanol (v/v) in ddH_2O. Continue with one bed volume n-butanol, 95, 67, 33% ethanol and at least five bed volumes of ddH_2O. Finish the regeneration by equilibration with starting buffer.

Store the gel in a refrigerator in the presence of 0.02% sodium azide.

References

Kennedy RM (1990) In: Deutscher MP (ed.) Guide to protein purification. Academic Press, San Diego, p 339

3.5.4 Analytical HPLC of Hapten-Protein Conjugates

Reversed-phase (rp-) or hydrophobic (HIC) HPLC is an excellent tool for analysis of proteins and peptides. Especially phenyl media, e.g., TosoHaas TSK-Gel Phenyl-5PW-RP, fit well

Solutions/Reagents

A 0.1% trifluoroacetic acid (TFA) (v/v) in ddH_2O

B 0.05% TFA (v/v) in acetonitrile (HPLC grade)

Use an HPLC apparatus with gradient system, pre-column and 5×0.5-cm separation column is used. Equilibrate the column with Soln. A at flow rate of 0.8 – 1.0 ml/min.

Filter the sample using a 0.45-μm syringe filter and fill a 20-μl sample loop. Inject and start the gradient program. The gradient program may be, for example, 0–30 min of 0–80% Soln. B in Soln. A, 5 min 100% Soln. B, 15 min 0% Soln. B (= 100% Soln. A). For enhanced resolution, steepness of the gradient and concentration range (%B) may be changed.

Serum albumin will be eluted at the given conditions at about 24 min. To detect very similar conjugation products modify the gradient: 0–10 min 0 → 30% B, 10–20 min 30 → 50% B, 30–40 min 50 → 80% B, 40–45 min 100% B, and 45–60 min 0% B. Monitor proteins at 225 nm or at the wavelength of the absorption maximum of the respective hapten.

Run the respective gradient program from time to time without protein to check for impurities. Inject about 2 ml 0.1 N NaOH to remove residual protein and to clean the columns and tubings.

3.6 Affinity Chromatography (AC)

Affinity chromatographic purification uses the more or less specific interactions between macromolecules or between macromolecules and low molar mass substances. The binding of a dissolved molecule (ligate) to an immobilized ligand should be not too strong, on one hand, because it has to be eluted without destruction, and on the other hand, it has to be very selective. Ligand-ligate pairs with very low dissociation constants K_D are not useful in affinity chromatography.

If proteins, e.g., antibodies, are used as immobilized ligands, medium ligand densities, e.g., 0.5–1 mg per milliliter of gel, give mostly better results than gels with a high amount of immobilized protein. Reasons for this observation might be steric hindrance between ligand molecules and/or between neighboring ligands and ligate as well as unspecific interactions between the ligand and non-ligate proteins of the sample. Tailing, i.e., the distance b of an elution peak is significantly larger than distance a (Fig. 3.5), or incomplete separations indicate such effects.

Immobilization via spacer

Since the binding sites of macromolecules are not always located on the surface, interactions between fixed ligands and these sites may be complicated or inhibited. Chemical modification of the ligand often overcomes these effects. So the introduction of a spatial distance between support surface and ligand by an inert spacer molecule often enables an binding of the ligate. But this spacer may also introduce hydrophobic or ion exchange effects or form clusters, which make interactions between ligate and ligand more difficult.

The first step in affinity chromatography is mostly the preparation of the affinity matrix by covalent immobilization (coupling)

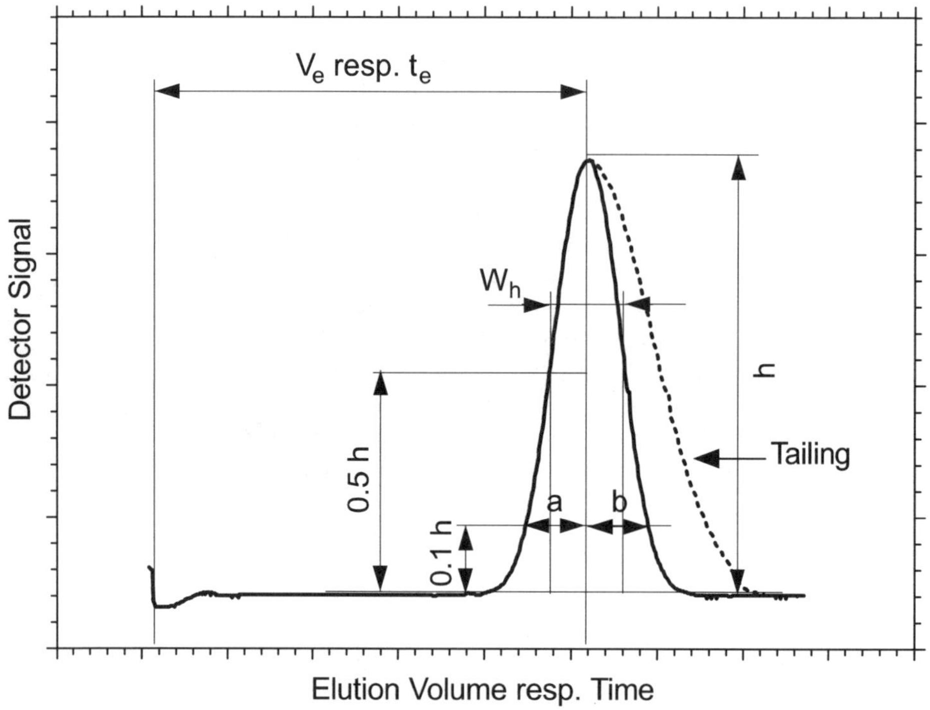

Fig. 3.5. Parameters of a signal peak obtained in chromatography. V_e elution volume, t_e elution time (retention time), h height of the peak, W width of the peak at 0.5 h, a and b indicators of peak symmetry

of the ligand to the carrier. Up to now, cyanogen bromide activation of polysaccharide carriers is the most used method for coupling, but since it has a relatively low chemical stability and is limited to amino groups bearing molecules, other coupling reagents should be taken into consideration. Table 3.6 gives a survey of coupling reagents useful in preparing affinity media. Table 3.7 lists reagents suitable for coupling ligands to spacer molecules, such as, for example, α, ω-diamines as 1,6-diaminohexane or spermidine or ω-aminocarbonic acids as 6-aminohexanoic acid (ε-aminocaproic acid).

Generally, sample application is done as described in Sect. 3.4.1. There are no differences between batch and column application and there are also no limits concerning volume as long as capacity of the matrix is not exceeded. Binding buffer should be optimal for biospecific interactions, i.e., pH, ionic strength, and composition should be optimized, and cofactors and structure-stabilizing agents should be present. Do not exceed protein concentration over 10 mg/ml and dilute sample or matrix if the affinity or ligand density is too high.

Elution of the bound ligand can be done by competition with the free (unbound) ligand or by partial denaturation of the pro-

Table 3.6. Reagents for direct immobilization of ligands

Reagent suitable for:	Proteins	Nucleic acids	$-NH_2$	-SH	Polysac-charides	at pH
Bisoxiranes	x		x	x	x	8.5–12
Bromocyan	x	x	x			9–11
Chloroformic acid esters	x	x	x	(x)		6–10
Divinylsulfone	x	x	x	x	x	8–10
Epichlorhydrine			x	x	x	8.5–12
Glutaraldehyde	x		x			6.5–9
Maleimide	x			x		6.5–7.5
N-Chlorcar-bonyloxy-2,3-di-carboximido-5-norbornene	x		x	(x)		4–10
Periodate	x	x	x			7.5–9
Trichloro-s-triazin	x	x	x		x	7.5–9

Table 3.7. Reagents for coupling of ligands to spacers

Reagent	Needed group on: Spacer	Ligand
Carbodiimide[a]	$-NH_2$	-COOH
Chloroformic acid esters[b]	-COOH	$-NH_2$, -OH
Maleimid-NHS derivative	$-NH_2$	-SH
N-Bromosuccinimide (NBS)	-COOH	$-NH_2$
N-Hydroxysuccinimide (NHS)[c]	-COOH	$-NH_2$
Orthoformic acid esters	$-NH_2$	-COOH

[a] e.g., EDAC (EDC)
[b] e.g., SMC
[c] e.g., ClCO-ONB

tein. The first case is demonstrated by the protocol for horseradish peroxidase purification on ConA Sepharose (Protocol 3.6.2.4), and the latter is illustrated by the purification of rabbit immunoglobulin A on protein A Sepharose (Protocol 3.6.2.6).

Competitive desorption (elution) has the advantage of very smooth conditions for protein structure, but the disadvantage of difficult removing of the ligand when further biospecific interactions are intended.

Denaturation of one or both partners of the affinity pair is done either by changing pH, by salts, or by chaotropic substances. This kind of desorption is easy to do, but there is danger of irreversible

Table 3.8. Buffers for elution in immunoaffinity chromatography

Buffer	pH
pH change	
0.1 M glycine-HCl	1.5 – 2.8
0.1 M glycine-HCl, 0.5 M NaCl	2.5
0.1 M Na acetate, 0.15 M NaCl, HCl	1
0.1 M Na citrate/phosphate	3.0 – 3.5
0.1 M acetic acid/formic acid	2.2
0.5 M acetic acid	
1 M propionic acid	
0.15 M NaCl/NH_4OH	11.0
0.1 M triethylamine	11.5
Chaotropic elution	
2.5 M KSCN	Unbuffered
0.5 – 3 M NaSCN	Unbuffered
3 M guanidinium hydrochloride	Unbuffered
6 M urea	Unbuffered
10% 1,4-dioxane (v/v)	Unbuffered
80% ethylenglycol (v/v)	Unbuffered
High ionic strength	
6.8 M NaCl	Unbuffered
4 M KI, 20 mM Tris-HCl	8.0
4 M $MgCl_2$, 10 mM sodium phosphate	7.0
5 M LiCl, 10 mM sodium phosphate	7.0

denaturation and/or precipitation. When denaturation is used for elution, the time in which the protein is under harsh conditions should be as short as possible. Some buffers for denaturation elution are given in Table 3.8.

References

Amersham Biosciences (2002) Affinity chromatography: principles and methods. www.chromatography.amershambiosciences.com

Dean PDG, Johnson WS, Middle FA (eds.) (1985) Affinity chromatography: a practical approach, IRL Press, Oxford

Jack GW (1994) Molec Biotechnol 1:59

Mohr P, Holtzhauer M, Kaiser G (1992) Immunosorption techniques: fundamentals and applications. Akademie Verlag, Berlin

Wilchek M, Miron T, Kohn J (1984) Meth Enzymol 104:3

3.6.1 Cyanogen Bromide Activation of Polysaccharide-Based Supports

***Caution!** Cyanogen bromide (BrCN, bromocyan) is extremely toxic. Work in a hood. Deactivate cyanogen bromide-containing solutions thoroughly before wasting.*

Cyanogen bromide is a solid with a high vapor pressure at room temperature; therefore, it is recommended to use a commercially available solution of BrCN in acetonitrile.

Equipment which was in contact with cyanogen bromide and washings has to be decontaminated with alkaline potassium permanganate solution before cleaning and disposal.

Solutions/Reagents

A 2 M Na_2CO_3 in de-ionized water
B 5.0 M BrCN in acetonitrile

This protocol describes the activation of an agarose gel, e.g., Sepharose 4B. Other polysaccharide-based supports are handled analogously.

Wash the agarose gel on a sintered-glass filter with about 5 vol. of de-ionized water. Suck off remaining water until a solid but not dry cake is obtained. Mix 10 g of this gel with 20 ml of Soln. A and cool in an ice bath to 0 °C. Do not use a magnetic stirrer!

In a hood add 2 ml Soln. B quickly to the ice-cold suspension, mix, suck off after 2 min, and wash with 150 – 200 ml ice-cold water. For a high degree of activation it is important to maintain the temperature low and to work quickly.

Use the activated gel immediately for coupling (e.g., Protocol 3.6.2), because the reactive groups hydrolyze fast.

As a variant use 1-cyan-4-dimetylaminopyridinium-tetrafluoroborate (CDAP) instead of BrCN. This reagent is not so hazardous and the degree of activation is easier to control.

Solutions/Reagents

A 30% acetone (v/v) in de-ionized water
B 60% acetone (v/v) in de-ionized water
C 30% acetone (v/v) in de-ionized water
D acetone/0.1 N HCl 1:1 (v/v)

Ten grams of Sepharose are sucked off on a sintered-glass funnel. First wash with 3 vol. water, then with 3 vol. 30% acetone in water (v/v), and then with 3 vol. 60% acetone (v/v) in water. Suspend the gel in 10 ml of 60% acetone and cool to 0 °C. Add the needed amount CDAP dropwise (cf. Table 3.9) followed by the respective amount of triethylamine (TEA) to the stirred gel suspension. Suck off the gel after 2 min and wash with a tenfold volume of Soln. D. The activated gel is stable for 1 h.

Table 3.9. Amount of CDAP and TEA for activation of 10 g wet Sepharose 4B

Degree of activation	Capacity (µMoles act. groups per gram)	mg CDAP	ml TEA
Low	5	15	0.2
Medium	15	75	0.6
High	30	150	1.2

3.6.1.1 Determination of the Degree of Activation

Solutions/Reagents

A 30 mg N,N′-dimethylbarbituric acid dissolved in 1.8 ml reagent-grade pyridine and made with ddH_2O up to 2.0 ml (some days stable at −20 °C)

Weigh in 5 mg of sucked-off activated gel into a test tube and add 200 µl Soln. A. Close the tube and mix for 15 min at RT. Add 2.80 ml ddH_2O and allow to settle or centrifuge. Read the supernatant at 588 nm in a photometer. If necessary, dilute with water.

Molar absorptivity of the reactive cyanate groups $\varepsilon_{1\,\mathrm{cm}} = 137\ \mathrm{ml}\ \mu\mathrm{Mol}^{-1}\ \mathrm{cm}^{-1}$.

Degree of activation is calculated by the following equation.

$$a = \frac{\mathrm{O.D.}_{588} \cdot 1000}{0.137 \cdot 3 \cdot \text{mg weighed portion}}$$

References

Kohn J, Wilchek M (1983) FEBS Lett 154:209
Kohn J, Wilchek M (1984) Appl Biochem Biotechnol 9:285
Porath J (1974) Meth Enzymol 34:13

3.6.2 Coupling to Cyanogen Bromide-Activated Gels

Forming of covalent binding (coupling) between activated gel and ligand needs free primary amino groups of the molecule to be immobilized (ligand). The ligand may be bound directly to the matrix or via an intervening molecule, the spacer.

To immobilize a protein the biologic activity of which depends on a cofactor, this cofactor has to be present in sufficient concentration during the immobilization step. Thus, for example, the lectin Concanavalin A is immobilized in the presence of a buffer containing 0.1 M glucose or α-methylmannoside, 1 mM calcium chloride and 1 mM manganese chloride.

Important! *The active center of a protein and the binding site to its ligate has to be protected by substrate, cofactor, ligand, or their analogue during coupling to the chromatographic support.*

Solutions/Reagents

A 0.5 M NaCl in 0.1 M borate, Tricine, TAPS, or carbonate buffer, pH 9.5
B 0.5 M NaCl in 0.1 M acetate buffer, pH 4.5
C 1 M NaCl in pure water
D 1 M ethanolamine hydrochloride, pH 8.0, in pure water

Wash freshly activated gel (see Protocol 3.6.1) with huge amounts of ice-cold water or in the case of commercially available activated gels, e.g., BrCN-activated Sepharose 6B, with ice-cold 0.001 N HCl.

Dissolve the protein (up to 10 mg/ml) or the spacer, e.g., 0.2 mg/ml 1,6-diamminohexane, in buffer A. Mix gel and protein solution in a ratio of 1 g wet gel to 1 ml solution in a flask and incubate at RT for 2 h or at 4 °C overnight. Pour the gel on a sintered-glass funnel and suck off the liquid. Bring the gel back to the flask and incubate with an equal volume of Soln. D for 2 h at RT (blocking remaining active groups). Gently agitate the gel during these steps.

After the second incubation wash the gel alternating with the 50-fold volume of Soln. A and B, at least twice. Bring the gel to neutral pH by rinsing with Soln. C.

Store the obtained affinity matrix in a buffer stabilizing the ligand properties, supplement with a biocide, e.g., 0.02% sodium azide or 0.1% Thimerosal or some drops of chloroform.

If a spacer such as, for example, 1,6-diaminohexane, was bound to the matrix, in a second step the carboxyl-groups-bearing ligand will be immobilized. For ligands with free amino groups the spacer could be 6-aminohexanoic acid. Coupling reagents often are water-soluble carbodiimides as N-ethyl-N′-(3-dimethylaminopropyl)-carbodiimide (EDC) or N-cyclohexyl-N′-[2-(4-methylmorpholinium)-ethyl]-carbodiimide tosylate (CMC) in slightly acidic buffers, e.g., 0.25 mm MES or phosphate, pH 4.5.

Water-soluble carbodiimides

If the ligand or the coupling reagent is not water soluble, couple in 80% dimethylformamide (DMF) (v/v) or 80% dimethylsulfoxide (DMSO) (v/v), adjusted to pH 4.5.

3.6.2.1 Quantitative Determination of Coupled Diamine Spacers with 2,4,6-Trinitrobenzene Sulfonic Acid

This is a modification of the TNBS procedure by Antoni et al. It is suitable for polysaccharide media hydrolysable by boiling glacial acetic acid.

Solutions/Reagents

A 0.5 M NaCl in 0.1 M borate buffer, pH 9.5
2,4,6-trinitrobenzenesulfonic acid
glacial acetic acid

Suspend 100.0 mg wet gel and 15 mg 2,4,6-trinitrobenzensulfonic acid in 5 ml Soln. A. Centrifuge with 3000 × g after 2 h at RT. Wash the pellet several times with ddH_2O by suspending and centrifugation.

Add 10 ml glacial acetic acid to the washed pellet and boil to reflux for 2 h. Cool to RT and determine the volume. Read absorbance at 340 nm.

Molar absorption coefficient of the formed sulfonamide $\varepsilon_{340} = 1.24 \times 10^4\ M^{-1}\ cm^{-1}$; specific gravity of wet Sepharose $\varrho = 1.1$ g/ml.

Bound amino spacer or immobilized protein may be determined also by the KJELDAHL method (Protocol 1.1.6).

References

Antoni G, Presentini R, Neri D (1983) Anal Biochem 129:60

3.6.2.2 Quantitative Determination of Immobilized Protein

The BCA assay for immobilized protein is more reliable than calculations from UV measurements of coupling supernatant.

Solutions/Reagents

- A 141 mM Na_2CO_3, 113 mM $NaHCO_3$, 100 mM NaOH, 7 mM sodium tartrate, pH 11.25
- B 1% BCA (w/v), disodium salt, in ddH_2O
- C 4% $CuSO_4{\cdot}5H_2O$ (w/v) in ddH_2O
- D 25 vol. Soln. A, 24 vol. Soln. B, and 1 vol. Soln. C; mix prior to use

Standard 1 mg/ml BSA in Soln. A

Wash the affinity matrix after coupling and blocking as described in Protocol 3.6.2. In a second step, wash about 50 µl of the support with 1 ml of Soln. A. Remove excess solvent by spotting onto filter paper (avoid drying).

Weigh triplicates of 5 mg of wet support into 1.5-ml reaction tubes, add 500 µl of Soln. A followed by 500 µl of Soln. D[2]. Shake at 60 °C for 1 h, and then read absorbance at 562 nm against water. The standard curve is made from 0 to 50 µg BSA in Soln. A.

References

Kakabakos SE, Tyllianakis PE, Evangelatos GP, Ithakissios DS (1994) Biomaterials 15:289

3.6.2.3 Immobilization of Wheat Germ Agglutinin

Solutions/Reagents

- A 0.1 M N-acetyl-glucosamine (GlcNAc, M_r 221.2), 1 M NaCl, 0.1 M borate buffer, pH 8.3
- B 0.5 M ethanolamine (3 ml 2-aminoethanol in 80 ml pure water, adjust pH to 8.0 with HCl, bring up to 100 ml with pure water)
- C 0.5 M NaCl in 0.1 M carbonate buffer, pH 9.0
- D 0.5 M NaCl in 0.1 M acetate buffer, pH 4.5
- E 0.9% NaCl (w/v), 0.02% sodium ethylmercurithiosalicylate (Thimerosal) in ddH_2O

[2] The same volumes are used when commercially available BCA reagents are used.

Dissolve in a centrifugation tube 10 mg wheat germ agglutinin (WGA) in 5 ml of Soln. A. Add 5 g of wet, ice-cold activated gel (from BrCN activation; cf. Protocol 3.6.1) and agitate the mixture at a roller shaker for 2 h at RT or overnight at 4 °C.

Important! *Don't use a magnetic stirrer bar.*

After incubation, add 10 ml Soln. B and block at RT further 2 h. Wash the gel consecutively with 50 ml each of Soln. C and Soln. D and repeat this step. Finish coupling by washing on a filter funnel with 100 ml of Soln. E, suck off the buffer and resuspend then in 5 ml of Soln. E. Store the gel in a refrigerator.

Testing capacity

The coupling yield is about 60%. If necessary, check the binding capacity by adsorption of fetuin and elution with 0.1 M N-acetyl-glucosamine in PBS.

The immobilization of any other protein to cyanogen bromide-activated or N-hydroxysuccinimide- (NHS) activated gels is done the same way.

3.6.2.4 Affinity Purification of HRP

Solutions/Reagents

A PBS

B 0.1 M α-methyl-D-glucoside or -mannoside in PBS

C 1 M NaCl, 1 mM $CaCl_2$, 1 mM $mgCl_2$, 0.02% Thimerosal (w/v) in 0.1 M acetate buffer, pH 6.0, supplemented with 1 mM $MnCl_2$ after adjusting pH

Dissolve the crude horseradish peroxidase (HRP) to about 10 mg/ml in Soln. A and centrifuge.

Wash a column with 5 ml Concanavalin A bed (1 × 6.4 cm) with 50 ml of Soln. A at RT. Apply up to 20 ml of the above HRP solution. When all of the dark brown solution is within the gel, wash with Soln. A with a flow rate of about 1 ml/min at RT, until the eluate is colorless. If the capacity of the column is not sufficient, the eluate may be rechromatographed after regeneration of the column.

Elute the bound colored HRP with Soln. B at 1 ml/min.

Regenerate the column with 25 ml of Soln. C and store it in a refrigerator until reusing.

3.6.2.5 Affinity Chromatography of Immunoglobulins on Immobilized Antibodies (Immunoaffinity Chromatography, IAC)

Solutions/Reagents

A PBS

B 0.1 M glycine-HCl buffer, pH 2.5

C 0.5 M phosphate buffer, pH 7.5, or 0.5 M Tris buffer pH 8.0

BrCN-immobilized IgG

Dilute the respective antiserum, e.g., goat anti-(rabbit IgG) anti-serum, to about 10 mg/ml with Soln. A and centrifuge or filter through a 0.45-µm membrane filter.

Apply 1 ml of the clear solution on top of a column containing the immobilized antigen, e.g., rabbit IgG. The bed size depends on the capacity of the support; in this example it should be 4–5 ml. Wash with Soln. A until the $O.D._{280}$ of the eluate is < 0.05.

Pipet 0.2 ml of Soln. C into tubes for fraction collection. If chromatography is performed in the cold, the phosphate buffer Soln. C tends to crystallize; therefore, add phosphate buffer immediately before use.

Elute the bound immunoglobulin with Soln. C at a flow rate of 1 ml/min. Mix the eluate thoroughly with the buffer within the collector tubes and monitor the elution by measuring UV absorption at 280 nm.

Combine fractions with $A_{280} > 0.4$, dialyze against the 100-fold volume PBS for at least 2 h, and centrifuge with $2000 \times g$ for 20 min. If necessary, concentrate the dialysate by ultrafiltration.

A milder elution is possible using 2 M $MgCl_2$, pH 5.0 instead of Soln. B. In this case Soln. C is omitted. For other conditions of elution see Table 3.7.

Especially if samples with a high content of unspecific proteins, e.g., sera, are processed by immunoaffinity chromatography, it is recommended to use a sequential washing/elution protocol: When all sample is applied, wash the column with PBS or TBS until $O.D._{280}$ < 0.1, then with 1 column volume 1 M sodium chloride, followed by 5 vol. PBS or TBS. Elute with 1 vol. elution buffer, e.g., glycine-HCl pH 2.5, then apply 1 vol. alkaline buffer, e.g., 0.1 M borate pH 10. Finally, regenerate the column with PBS or TBS.

References

Jack GW (1994) Molec Biotechnol 1:59

3.6.2.6 Affinity Chromatography of Rabbit IgG on Protein-A Supports

The specificity of protein A for immunoglobulins of different species and subclasses is given in Table 4.6.

Solutions/Reagents

- A PBS, pH 7.8
- B 0.5 M NaCl, 50 mM sodium phosphate, pH 7.8
- C 0.1 M sodium citrate, pH 3.0[3]
- D 1 N NaOH

Rinse a protein A column (4–5 ml bed volume; 15 × 22.5 and 15 × 28.5 mm, respectively) with 40–50 ml of Soln. A, flow rate 1 ml/min. Dilute 2–5 ml of rabbit antiserum with the same volume of Soln. A and apply with the same flow rate to the column. After sample

[3] Since the immunoglobulins purified by this method are mostly used for immobilization, buffers containing NH_2 groups, as Tris or glycine, should be avoided.

loading, wash the column with Soln. B until the UV absorption at 280 nm is below 0.01. Elute the bound immunoglobulin with Soln. C. Fractions giving absorption $A_{280}^{1\,cm} > 0.2$ are collected and immediately neutralized with Soln. D. Centrifuge the combined eluates to remove aggregates. If necessary, dialyze the eluate three times against a 20-fold volume of PBS for 2 h each.

Regenerate the column with 60–75 ml of Soln. A.

Determine the immunoglobulin concentration of the eluate photometrically (cf. equation d), Chap. 1.1.7).

Store the protein solution either in the presence of a biocide as sodium azide or Thimerosal a 4 °C or add glycerol to a final concentration of 30–50% (w/v) and store at −20 °C.

References

Harlow E, Lane D (1988) Antibodies. A laboratory manual. Cold Spring Harbor Laboratory. p 310

3.6.3 Activation of Sepharose with Epichlorohydrin

Epoxy-activated gels are especially used for immobilization of carbohydrates.

Solutions/Reagents

A 1 N NaOH

Mix 30 g of wet Sepharose CL-4B with 40 ml of Soln. A, 5 ml epichlorohydrin and 25 ml pure water. Incubate at 40 °C with gentle agitation for 2 h. Then wash on a sintered-glass funnel with 500 ml water and suck off remaining water. Activated gel is stable at 4 °C for several days.

***Caution!** Epichlorohydrin is toxic. Use a hood!*

3.6.3.1 Determination of Epoxy Residues

Solutions/Reagents

A 1.3 M $Na_2S_2O_3$ (3.226 g $Na_2S_2O_3 \cdot 5H_2O$ in 10 ml pure water, adjusted to pH 7.5)

B 10 mM HCl

Add 3.0 ml of Soln. A to 1.0 g of wet gel. Slurry the mixture with 20 ml ddH_2O after 30 min at RT and titrate with Soln. B to pH 7.0. Use a glass pH electrode for monitoring of pH.

One milliliter of used solution is equivalent to 10 µMoles of epoxy groups. After determination of the specific gravity of wet gel calculate the content.

3.6.4 Immobilization of Monosaccharides (Fucose)

Solutions/Reagents

A 1 N NaOH

fucose

ethanolamine

Shake 15 g of the washed wet Sepharose together with 10 ml of Soln. A, 0.5 g fucose, and 75 ml pure water at 40 °C overnight. Add 0.6 g of ethanolamine and continue shaking at 40 °C for further 2 h. Remove liquid and wash with 500 ml pure water.

References

Matsumoto I, Mizuno Y, Seno N (1979) J Biochem 85:1091
Sundberg L, Porath J (1974) J Chromatogr 90:87

3.6.5 Activation with Divinylsulfone

Carbohydrates as well as proteins are effectively coupled to divinylsulfone-activated agaroses. This variant is often more favorable then the activation by epoxides. As further advantage, the resulting vinyl group serves as a spacer.

Solutions/Reagents

A 0.5 M Na_2CO_3, adjusted to pH 11 with NaOH
B 0.5 M Na_2CO_3, adjusted to pH 10 with NaOH
C 0.1 M ethanolamine, adjusted to pH 8 with HCl

Wash 25 ml settled gel (e.g., Sepharose CL-4B) on a Büchner funnel with 100 ml pure water, followed by 50 ml Soln. A. Transfer the gel into a lockable flask and add 25 ml of Soln. A. Resuspend the gel and add 2.5 ml of divinylsulfone. Close the flask and agitate at RT for 1–2 h. Next wash the gel on a Büchner or sintered-glass funnel with 200–250 ml of Soln. B. Suck off the excess liquid.

Transfer the wet gel into a bottle (e.g., 50-ml Falcon tube), and add the ligand dissolved in 25 ml of Soln. B. 0.5–1 mmoles or 5–10 mg of protein per milliliter are optimal. Less ligand is also possible. If only few ligand is available, reduce the amount of gel and buffer.

Agitate the mixture on a roller at RT for 1–3 days. Next wash the gel with 200–250 ml of Soln. B on a sintered-glass funnel.

Block remaining active groups by incubation with 25 ml of Soln. C at 4 °C for 3 h, suck off, and equilibrate the gel with that buffer which will be used in affinity chromatography.

Store the gel in a refrigerator in the presence of 0.02% NaN_3 or Thimerosal.

References

Ultrogel, Magnogel, and Trisacryl: Practical guide in affinity chromatogaphy and related techniques, 2nd ed. (1983) Réactifs IBF, Villeneuve-la-Garenne

3.6.6 Coupling of Reactive Dyes to Polysaccharides (Dye-Ligand Chromatography)

Immobilized reactive dyes serve as ligands in affinity chromatography. They seem to be structural analogs of cofactors and therefore the selectivity of dye-ligand support is rather high. Nevertheless, a prediction as to which dye is the most suitable for separation is very difficult and the optimal dye has to be chosen experimentally. For this purpose, several suppliers offer kits containing several dyes. The procedure for testing binding and elution as well as capacity is similar to IEC and HIC.

References

Stellwagen D (1990) In: Deutscher MP (ed.) Guide to protein purification. Meth Enzymol 182:343

The procedure given in this protocol for coupling of reactive dyes (triazine dyes), such as, for example, Cibacron Blue F3G-A (C.I. 61211) or Reactive Red 4 (C.I. 18105), is applicable to most polysaccharide beds.

Suspend 10 g dry or 300 ml swollen beads in a total volume of 350 ml water, stir with a moving blade agitator, and heat to 60 °C. Now add dropwise a solution of 2 g dye in 60 ml water. Add 45 g of sodium chloride 30 min after finishing the addition of dye and continue stirring at 60 °C for an additional hour. Next raise the temperature to 80 °C and add 4 g of anhydrous sodium carbonate. Continue stirring at 80 °C for 2 h.

Wash the gel after cooling to RT with pure water until the filtrate is colorless.

The gels are resistant especially to high concentrations of urea or thiocyanate and they are stored as usual in a refrigerator in the presence of a biocide.

References

Böhme H-J, Kopperschläger G, Schulz J, Hofmann E (1972) J Chromatogr 69:209

3.6.7 Covalent Coupling of Biotin (Biotin-Avidin/Streptavidin System)

The binding of biotin to the tetramere proteins avidin and streptavidin is very specific with an extremely low dissociation constant ($K_D = 10^{-15}$ M). Because each of the subunits of the proteins carry a binding site, it is possible to produce larger complexes when derivatives are used bearing at least two biotin molecules.

Avidin from hen egg is a glycoprotein, which is easily conjugated to other proteins by periodate activation (cf. Protocol 4.1.6).

For affinity chromatography, avidin or streptavidin are coupled to supports using Protocol 3.6.2. Biotin is covalently bound to proteins or other primary amino groups bearing molecules using its N-hydroxsuccinimide ester[4].

Not only biotinylated proteins, but also nucleic acids or peptides synthesized in the presence of biotin derivatives or carbohydrate-biotin conjugates, are separated by immobilized avidin and streptavidin, respectively, or are identified by avidin-enzyme or avidin-gold conjugates.

Mostly the accessibility of an immobilized ligand to its ligate is increased if the ligand is bound to the matrix via a spacer. If steric hindrance is suspected, use a lengthened biotin derivative as N-6-biotinylamidohexanoyl hydroxysuccinimide (Biotin-LC-NHS).

The following protocol is an example for the biotinylation of immunoglobulins in a molar ratio of 1 mole IgG to 5 moles biotin. Depending on the kind of protein, ratios up to 1:20 are possible.

Because at a given pH the reactivity and/or accessibility of NH_2 groups of amino acid side chains of a protein are not the same, the number of conjugated biotin residues may be altered by variation of the buffer pH between pH 6 and 9 (the higher pH the larger the degree of conjugation).

Solutions/Reagents

A PBS or saline

B 1 mg biotinyl-N-hydroxysuccinimide ester[5] (BNHS; 58.6 n Moles/μl, M_r 341.4) in 50 μl DMF (ultrapure or distilled), prepare freshly

Supplement 250 μl of antibody solution (about 2 mg of antibody per milliliter of Soln. A) with 1.1 μl of Soln. B and shake protected from light at RT for 1 h. Dialyze the reaction mixture three times with a 100-fold volume of Soln. A at 4 °C for 6 h each.

Add an equal volume of glycerol to the dialyzed sample and store at −20 °C.

Other proteins, such as, for instance, horseradish peroxidase or alkaline phosphatase, are labeled (conjugated) in the same way.

References

Bayer EA, Wilchek M (1980) In: Glick D (ed.) Methods of biochemical analysis, vol. 26. Wiley, New York, p 1

Savage MD, Mattson G, Desai S, Nielander GW, Morgensen S, Conklin EJ (1992) Avidin-biotin chemistry: a handbook. Pierce Chem. Comp., Rockford, Illinois, p 273

Hermanson GT (1996) Bioconjugate techniques. Academic Press, San Diego, p 377

[4] The preparation of NHS-biotin is described in: Bayer E, Wilchek M (1974) Meth Enzymol 34:265

[5] Instead of BNHS the more hydrophilic derivative sulfo-NHS-biotin (M_r 443.42) or a spacer derivative, e.g., sulfo-NHS-LC-biotin may be used.

3.6.8 Metal Chelate Chromatography of Proteins Containing His_6-Tag

For identification and purification, recombinant proteins are often tagged to the N-terminus with an additional sequence of histidyl residues, mostly six (His_6 tag). This tag binds selectively to cations as nickel or copper immobilized by covalent chelators as nitrilotriacetic acid. The method is named Metal Chelate Chromatography (MCC, MCAC, IMAC).

If the His tag interferes during further experiments, it may be cleaved enzymatically with specific endo- or exoproteases.

Solutions/Reagents

A 0.1 M $NiSO_4$ in ddH_2O
B 20 mM NaH_2PO_4, 0.5 M NaCl, 4 M guanidinium hydrochloride[6],
1 mM β-mercaptoethanol or DTT, 5 mM histidine, pH 8.0[7] [8]
C 20 mM NaH_2PO_4, 0.5 M NaCl, 4 M guanidinium hydrochloride,
1 mM β-mercaptoethanol or DTT, 20 mM histidine, pH 8.0
D 20 mM NaH_2PO_4, 0.5 M NaCl, 4 M guanidinium hydrochloride,
1 mM β-mercaptoethanol or DTT, 300 mM histidine, pH 8.0
E 20 mM NaH_2PO_4, 0.5 M NaCl, 300 mM histidine, pH 8.0

Preparation of the column: Wash the 1-ml bed (e.g., Chelating Sepharose (Ni Sepharose, "NTA resin") with 5 ml ddH_2O, followed by 0.5 ml of Soln. A, 5 ml ddH_2O, and 10 ml of binding buffer Soln. B at a flow rate of up to 1 ml/min.

Isolation of protein: Apply up to 1 ml sample. Wash with 10 ml of Soln. C and elute the bound protein with 2 ml of Soln. D. Monitor the elution by measuring A_{280} and check the fractions by dot blot using a mAb against His_6.

For the next chromatography, equilibrate the column with 10 ml of Soln. B. For storage of the column, wash with 10 ml ddH_2O followed by 10 ml 20% ethanol (v/v) in ddH_2O.

If the sample was denatured by extraction from cell lysates, a renaturation is possible while the protein is bound to the column: Wash the column with Soln. B and elute with a gradient formed by Soln. C (starting buffer) and Soln. E.

References

Ostrove S, Weiss S (1990) Meth Enzymol 182:371

[6] Instead of 4 M guanidinium hydrochloride 6 M urea may be used.
[7] Adjust pH of the ready mixed buffer with NaOH.
[8] For a non-denaturating preparation take Soln. B-D without guanidinium hydrochloride and urea, respectively, and β-mercaptoethanol and DTT, respectively.

Amersham Biosiences Application note: HiTrap chelating. http://www.amershambiosciences.com
Qiagen TAGZyme^{TM} Handbook for exoproteolytic cleavage of N-terminal His tags. http://www.qiagen.com

3.7 Concentration of Diluted Protein Solutions

3.7.1 Acidic Precipitation

Proteins in aqueous samples with concentrations of at least 5 µg/ml are precipitated by acids as trichloroacetic acid (TCA) or sulfosalicylic acid. The final concentration of acid should be 7.5 – 10% (w/v). Biologic activity is mostly and antigenicity is sometimes destroyed by acidic precipitation.

Solutions/Reagents
A 100% TCA (w/v) in ddH_2O
B 10% TCA (w/v) in ddH_2O
C 0.15% sodium deoxycholate (NaDOC) (w/v) in ddH_2O

Cool the sample in an ice bath and add 1/10 volume of Soln. A. Vortex and allow to precipitate in an ice bath for 30 min. Spin for 5 min at 5000 to 10 000 × g in a refrigerated centrifuge. Remove the supernatant by aspiration with a pipet. Wash the pellet once with 0.5 sample volume ice-cold 10% TCA (w/v) and once with 0.5 sample volume ice-cold ethanol-ether 1:1 (v/v) to remove traces of TCA. Dry the pellet at the air and dissolve in 0.1 N NaOH or 0.1 M Tris pH 8.0.

Using NaDOC, the protein concentration precipitable by TCA is below 1 µg/ml. The precipitation protocol is modified as follows: Add 0.1 ml Soln. C per milliliter to a sample. Mix and incubate at RT for 10 min. Add 0.1 ml of Soln. A, cool to 0 °C and continue as described above. In the presence of SDS the NaDOC-TCA precipitation does not work.

Proteins from tissue homogenates are precipitated by tungstic acid. Mix homogenates, which contain about 1% protein, with 1/10 of volume of 10% sodium tungstate ($Na_2WO_4 \cdot H_2O$) (w/v), and acidify with the same volume of 0.67 N sulfuric acid. Collect the precipitate after 10 min at RT by centrifugation. Nucleic acids are precipitated by tungstic acid to a minor extent.

References

Bollag DM, Edelstein SJ (1991) Protein methods. Wiley-Liss, New York

3.7.2 Salting Out

Proteins are salted out (precipitated) by ammonium or sodium sulfate (see Hofemeister series, Sect. 3.2). The structure of proteins is much better conserved by salt precipitation than by acidic precipitation. The salt concentration needed for precipitating a protein depends on its nature and on temperature; therefore, salting out may be used for fractionation of protein mixtures (cf. Protocol 4.3).

Fractionation conditions are expressed as "% saturation". Table 8.19 and 8.20 give the amount of solid ammonium sulfate necessary to get a distinct degree of saturation starting from a certain starting saturation.

To precipitate a protein, mix its solution either at RT or 4 °C with the required amount of saturated ammonium sulfate solution (consider the different concentration of $(NH_4)_2SO_4$ at RT and 4 °C). As an example, mix 6 vol. of protein solution with 4 vol. of saturated ammonium sulfate solution to get 40% saturation.

> ***Important!*** *Saturated ammonium sulfate has to have a solid salt within the flask. When preparing fresh solution, saturation equilibrium is reached after several hours.*

Allow the precipitate to form overnight at the required temperature (do not store in a refrigerator if "% saturation" is made for room temperature). Spin with 2000 × g for 10 min, wash the pellet by suspending in ammonium sulfate solution of the indicated degree of saturation, spin again and decant the supernatant. If the precipitate contains the wanted protein, solve it in pure water and dialyze against an appropriate buffer.

3.7.3 Precipitation Using Organic Substances

By addition of up to 80% (v/v) of organic solvents such as methanol, ethanol, or acetone to protein solutions, a formation of precipitates occurs. Varying the solvent concentration allows a fractionation as described for salting out. Because organic solvents tend to denature proteins at temperatures above 10 °C, precipitation and all further steps have to be performed at 0 °C or below. Some buffer salts may precipitate also at elevated concentrations of organic solvents; therefore, the ionic strength of buffers should not be above 0.2.

After collection of the precipitate by centrifugation, the pellet is washed and resuspended in a suitable buffer of appropriate volume. Insoluble, denatured material is removed by centrifugation or filtration.

Removal of SDS

A very special example of protein precipitation by organic solvents is the removal of SDS and other detergents from PAGE samples prior to protein determination: Mix 0.1 ml sample with 0.4 ml methanol and centrifuge at 9000 × g after 1 min. Add 0.1–0.2 ml chloroform and centrifuge again. Make a two-phase system by addition of 0.3 ml ddH_2O, centrifuge and collect the lower, organic phase together with the interphase zone. Add 0.3 ml methanol to the organic phase and centrifuge for 2–3 min. Aspirate the liquid and dry the pellet with air. After dissolving the pellet in an alkaline solvent, the sample may be used for protein determination.

References

Wessel D, Flügge UI (1983) Anal Biochem 138:141

Polyethylene glycol (PEG) is well suited for gentle concentration of proteins. Among the PEGs of different degree of polymerization, the most effective is PEG 6000. The concentration range of PEG 6000 is between 0 and 15% (w/v) for fractionation. As examples, circulating immune complexes are precipitated with 3.5% PEG 6000 (w/v, final concentration) in 0.1 M borate buffer, pH 8.4; membrane protein complexes are precipitated at 0 °C with 7.5% PEG 6000 (w/v, final concentration) in 50 mm Tris, 10 mm $mgCl_2$, pH 7.4.

Complete precipitation by PEG needs 30 min to 15 h at 0 °C. The precipitate is collected by centrifugation or filtration on glass fiber filters (e.g., Whatman GF/A). Dissolve the precipitate in ddH_2O and dialyze against an appropriate buffer. An increase in volume occurs during dialysis.

3.7.4 Lyophilization (Freeze Drying)

During lyophilization the water of deeply frozen samples is removed by sublimation in vacuum. The structure of a probe is well conserved, especially if additives as sucrose or trehalose are added. In a sealed bottle, the sample can be stored at RT for years. If the buffer constituents are to be removed also, the sample has to be dissolved in a volatile buffer (see Table 7.6). Reconstitute the sample by addition of water or, in the case of volatile buffers, by addition of a buffer of your choice.

For optimal conditions it is necessary to freeze the sample very quickly in a methanol-dry ice bath or in liquid nitrogen and to store the frozen sample below −30 °C until the vacuum of less than 100 Pa (0.0145 psi) is reached and the lyophilization is finished.

During the lyophilizing process, the sample has to get external heat because sublimation of ice withdraws heat resulting in cooling the probe, but avoid melting of the sample.

Concentration of a sample in a SpeedVac, a centrifuge with open rotor running in vacuum, is not lyophilization because the sample is not frozen; therefore, this method does not fit for macromolecules with defined structures. The reconstitution (dissolving) of protein samples concentrated in a SpeedVac is often more difficult and comparable to drying with air. But it is the method of choice for concentration of solutions of low-molar mass substances, e.g., peptides, since the sample is collected at the bottom of the tube by means of centrifugal force.

3.7.5 Ultrafiltration

Solutions of macromolecules may be concentrated by means of polymer membranes of defined pore size. Applying a pressure or centrifugal force, small molecules pass the pores, whereas large molecules retain. The nominal cutoff of an ultrafiltration membrane (MWCO) helps you to select a membrane: Molecules smaller than the MWCO will pass the membrane, whereas larger molecules are held back. This separation is not sharp and depends on protein conformation and solvent composition. Complete retention is achieved when using a membrane with a MWCO 1/3 to 1/5 of the molar mass of the macromolecule of interest. Figure 3.6 illustrates the separation of proteins by ultrafiltration.

Recovery of proteins from ultrafiltration depends not only on the size, but also on the solute composition and the type of membrane, since unspecific adsorption of the protein to the membrane cannot be excluded. Furthermore, the chemical resistance of the membrane to buffer components and sanitation ingredients should be taken into consideration.

Unspecific adsorption of proteins may be decreased if the membrane is preincubated with 5% Tween 20 (w/v) for 1 h and subsequently washed with ddH_2O.

Desalting is also possible by ultrafiltration. For this purpose, the sample is concentrated to about 10% of its volume, reconstituted to its original volume with water or a second buffer, and ultrafiltration is repeated twice in the same way. Proteins are less likely to be denatured, because ultrafiltration is a mechanical separation which does not need harsh chemicals for separation.

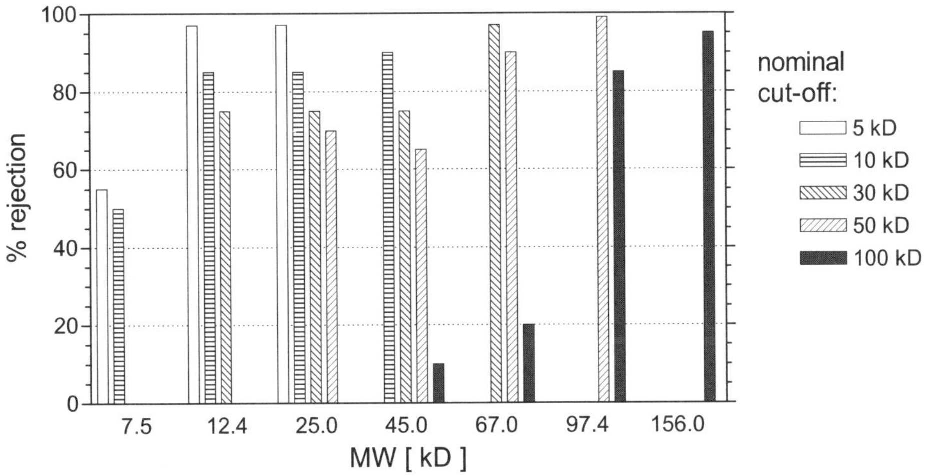

Fig. 3.6. Percentage of retention of proteins by ultrafiltration membranes. (Data according to Millipore/Amicon application note)

Aqueous samples may be concentrated using dialysis tubing. After dialysis, the tubing containing the sample is surrounded with dry GPC medium, e.g., Sephadex 200, or polyethylene glycol, e.g., PEG 20000 (if PEG is used, parts of the polymer with significantly lower molar mass may penetrate the membrane resulting in precipitation of the protein within the tubing). After several hours at 4 °C, water and buffer substances are sucked off and the volume of the dialysate is diminished. The buffer composition is not altered by this very smooth concentration method.

4 Immunochemical Protocols

The protocols given in this chapter are selected examples of the universe of modern immunology. The demands of a biochemical lab gave the background of this selection; therefore, procedures which afford specialized knowledge, such as, for example, production of monoclonal antibodies, are omitted, and the reader is referred to the respective literature.

As with other biochemical methods, the immunochemical protocols are improved, but modifications are possible and sometimes desirable.

References

Caponi L, Migliorini P (eds.) (1999) Antibody usage in the lab. Springer, Berlin

Crowther JR (2001) ELISA – Guidebook. Humana Press, Totowa, N.J.

Harlow E, Lane D (1988) Antibodies. A laboratory manual. Cold Spring Harbor Laboratory

Hudson L, Hay F (1989) Practical immunology, 3rd ed., Blackwell, Oxford

Kerr MA, Thorpe R (eds.) (1994) Immunochemistry Labfax. Academic Press, Oxford

Manson MM (ed.) (1992) Immunochemical protocols (Methods in molecular biology, vol. 10). Humana Press, Totowa N.J.

Mohr P, Holtzhauer M, Kaiser G (1992) Immunosorption techniques: fundamentals and applications. Akademie Verlag, Berlin

Peters JH, Baumgarten H (eds.) (1992) Monoclonal antibodies. Springer, Berlin

Phillips TM, Dickens BF (2000) Affinity and immunoaffinity purification protocols. Eaton Publishing, Natick

Schade R, Behn I, Erhard M, Hlinak A, Staak C (eds.) (2001) Chicken egg yolk antibodies, production and application. Springer, Berlin

Wild D (2005) The immunoassay handbook. Elsevier, Oxford

4.1 Conjugation of Haptens (Peptides) to Carrier Proteins

Low-molar mass molecules, such as, for example, chemically synthesized peptides, are in most cases not able to induce antibodies in animals; therefore, these molecules have to be covalently coupled (conjugated) to carrier molecules. Such macromolecular carriers are, for example, the hemolymph of the keyhole limpet *Megathura crenulata* (KLH), the hemolymph of the eatable snail *Helix pomatia*, the hemolymph of the horseshoe crab *Limulus polyphemus*, cationized serum albumin, hen egg ovalbumin, thyreoglobulin, or dextranes.

To get a directed coupling of a peptide via its N-terminal NH_2-group, the introduction of a monochloroacetylglycyl residue during peptide synthesis is recommended. Of course, coupling via NH_2-, SH-, or carboxyl groups of peptide side chains is also possible.

Production of antiserum with high titer and specifity is done by trial and error, especially because each immunized animal gives antisera with different characteristics; therefore, several groups of animals should be immunized with different antigen preparations.

Universal haptens, suitable especially in immunohistochemistry, dot blots, Western blots, or other kinds of immunoassays, are, for instance, biotin, fluorescein, or digoxigenin.

Some coupling reagents suitable for preparation of hapten-carrier conjugates are given in Table 4.1.

Table 4.1. Selected reagents for protein and peptide conjugation to carrier proteins

Coupling reagent	Peptide Binding to	Carrier protein
1-Cyclohexyl-3-(2-morpholin-4-yl-ethyl) carbodiimide	$-NH_2$, -COOH	-COOH, $-NH_2$
1-Ethyl-3-(3-dimethyl-aminopropyl) carbodiimide (EDC; EDAC)	$-NH_2$, -COOH	-COOH, $-NH_2$
2-Iminothiolane (Traut's Reagent)	$ClCH_2CO-NH-CH_2-CO-$	-SH
3-Maleimidobenzoic acid (MBS) N-hydroxysuccinimide ester	-SH, $-NH_2$	$-NH_2$, -SH
Succinimidyl 4-(N-maleimidomethyl)-cyclohexane-1-carboxylate (SMCC)	-SH, $-NH_2$	$-NH_2$, -SH
Aryldiazonium chloride	-SH, -Tyr	-SH, -Tyr
Bis-diazotated benzidine (BDB)	-Tyr	-Tyr
Glutardialdehyde	$-NH_2$	$-NH_2$
Sodium periodat	$-NH_2$	Saccharide side chain
N-Succinimidyl-3-(2-pyridyl-dithio) propionate (SPDP)	$-NH_2$	$-NH_2$
4-Azido-benzoic acid N-hydroxysuccinimide ester (HSAB)	(photoactivation)	$-NH_2$

Further coupling reagents are substances used in cross-linking proteins, which allow an estimation of interacting area between proteins or parts of a protein. Because of the chemical structure of these substances, spacers are introduced between hapten and carrier.

Cross-linking reagents may be divided into several types:

- Water soluble (contain mostly sulfonic acid residues) and water insoluble (penetrate mostly biological membranes)
- Homobifunctional (two identically reactive groups) and heterobifunctional (at least two chemically different reactive groups)
- Cleavable and not cleavable

Examples of cross-linking bifunctional agents are given in Table 4.2.

Table 4.2. Selected homo- and heterobifunctional cross-linking reagents

Reagent	Homo	Hetero	Water soluble	Cleavable	Distance (nm)
Glutaraldehyde	x		Yes	No	
Dimethyladipinimidate (DMA)	x		Yes	No	0.86
Disuccinimidylsuberate (DSS)	x		No	No	1.14
Bis[β-(4-azidosalicyl-amido)-ethyl]-disulfide (BASED)	x		No	Yes	3.47
Bis[2-(Sulfosuccinimido-oxycarbonyl-oxy)ethyl] sulfone (Sulfo-BSOCOES)	x		Yes	Yes	1.30
Disuccinimidyltartrate (DST)	x		Yes	Yes	0.64
Dimethyl-3,3′-dithiobispropionimidate hydrochloride (DTBP)	x		Yes	Yes	1.19
Succinimidyl 4-(N-maleimidomethyl)-cyclohexane-1-carboxylate (SMCC)		x	No	No	1.16
m-Maleimidobenzyl-N-hydroxysuccinimide ester (MBS)		x	No	No	0.99
N-Sucinimidyl-3-(2-pyridyldithio) propionate (SPDP)		x	No	Yes	0.68
4-Sucinimidyl-4-(p-maleinimidophenyl) butyrat (SMPT)		x	No	Yes	1.12

References

Aslam M, Dent AH (1998) Bioconjugation. Protein coupling techniques for the biomedical sciences. Macmillan Reference Ltd., London

Hermanson GT (1996) Bioconjugate techniques. Academic Press, San Diego

Hosoda H, Ishikawa E (1993) Coupling of haptens to carriers and to labels. In: Masseyeff RF, Albert WH, Staine NA (eds.) Methods of immunological analysis. Samples and reagents. VCH, Weinheim, p 465

Hudson L, Hay FC (1990) Practical immunology, 3rd ed. Blackwell, Oxford, p 44

Wong SS (1991) Chemistry of protein conjugation and cross-linking. CRC Press, Boca Raton

4.1.1 Activation of Proteins with Traut's Reagent Yielding Proteins with Additional Free SH Groups

Solutions/Reagents

A 0.9 M NaCl, 0.1 M $NaHCO_3$, pH 8.0

B 0.9 M NaCl, 10 mM EDTA, disodium salt, 50 mM sodium phosphate, 0.02% NaN_3, pH 7.4

C PBS, 10 mM EDTA, disodium salt, 0.02% NaN_3

Mix 4 mg of KLH or a respective amount of a glycerol-containing solution of KLH (volume maximal 60 µl) with 190 µl Soln. A. Weigh 0.6 mg Traut's reagent (2-iminothiolane) into an Eppendorf tube and add the carrier protein solution. Shake at RT for 30 min.

Equilibrate a 10-ml Sephadex G-25 column with Soln. B and determine the void volume using dextran blue. Apply the reaction mixture and elute with Soln. B. Use the fractions containing the activated protein (in the case of KLH a light blue solution) immediately for conjugation (e.g., Protocol 4.1.3).

If BSA or ovalbumin is used, dissolve and elute the protein in Soln. C and monitor elution by UV reading.

Conjugate MCA-Gly peptides or maleimid-activated proteins to the iminothiolane-acivated carriers.

4.1.2 Conjugation of MCA-Gly Peptides to SH-Carrying Proteins

Weigh 4 mg of a monochloroacetylglycyl peptide (MCA-Gly peptide, peptide carrying a MCA-glycyl residue at the N-terminus) into an Eppendorf tube and add the activated KLH or other iminothiolane-activated carrier protein. Shake vigorously at RT for 3 h. Dialyze the reaction mixture twice at RT against PBS for 1 h each. Calculate protein concentration from 235-, 260-, and 280-nm readings (cf. Protocol 1.1.7).

Filter the conjugate through a 0.22-µm-tip filter and store in a refrigerator until immunization. For boostering, aliquot the sterile filtrate and freeze at −70 °C or mix with glycerol 1:1 and store at −20 °C.

Important: Prepare conjugates with different carrier proteins for immunization and testing and use different coupling reagents, if possible.

References

Lindner W, Robey FA (1987) Int J Peptid Res 30:794

4.1.3 Conjugation of Sulfhydryl Peptides Using 4-(N-Maleimidomethyl)-Cyclohexane-1-Carbonic Acid N-Hydroxysuccinimide Ester (SMCC)

Solutions/Reagents

A 50 mM HEPES, pH 7.4[1]
B 50 mM HEPES, 10 mM EDTA, pH 6.8
C 0.1 M sodium phosphate, 0.15 M NaCl, pH 7.2

Dissolve 4 mg of the carrier protein in 400 µl Soln. A. Dissolve about 0.5 mg SMCC (succinimidyl 4-(N-maleimidomethyl)-cyclohexane-1-carboxylate, M_r 334.3)[2] in 50 µl DMF and add this solution to the carrier protein. Shake at RT for 1 h, centrifuge and desalt on a Sephadex G-25 column, equilibrated with Soln. B (the activated carrier appears in the void volume). Cool the receiving tube in an ice bath.

Dissolve 4 mg of a Cys-containing peptide in 400 µl Soln. B and mix with the solution of the activated carrier protein. Shake at RT for 1 h, aliquot and freeze at −70 °C.

Conjugation of antibodies

Dialyze an antibody solution against Soln. C and concentrate to 20–30 mg/ml. Add 3 mg of sulfo-SMCC (3-sulfosuccinimidyl-4-(N-maleimidomethyl)-cyclohexane-1-caboxylate, M_r 436.7) or of sulfo-GMBS (N-(γ-maleimidobutyryloxy)-3-sulfo-N-hydroxysuccinimide ester, M_r 382.3), dissolved in ddH_2O to 60 mg/ml, to 10 mg of antibody, rock at RT for 15 min and add further 3 mg of coupling reagent. Desalt on a Sephadex G-25 column, equilibrated with Soln. A, after total incubation time of 30 min and use the activated antibody (protein) for conjugation immediately (see above).

References

Hermanson G (1996) Bioconjugate techniques. Academic Press, San Diego. p 235, p 444

[1] If KLH is used, the buffer should contain 0.9 M NaCl and 0.25 mg SMCC/4 mg KLH are used.

[2] Instead of the DMF solution of the water-insoluble SMCC, the same amount of sulfo-SMCC in water my be used.

4.1.4 β-Galactosidase-Immunoglobulin Conjugate (Coupling via SH Groups)

β-Galactosidase from *Escherichia coli* contains sufficient SH groups for conjugation; therefore, only the antibody has to be activated by introducing maleimide groups which react with the SH groups of the enzyme. Conjugation of other proteins or haptens may be performed analogously.

Solutions/Reagents

A 0.1 M sodium phosphate, 0.15 M NaCl, 10 mM EDTA, pH 7.2
B 0.1 M sodium phosphate, 0.15 M NaCl, pH 7.2
C ***storage buffer tenfold***: 0.1 M sodium phosphate, 1 M NaCl, 10 mM magnesium acetate, 1% NaN_3 (w/v), 10 mg/ml BSA, pH 6.5

Dissolve the enzyme to a concentration of 1–2 mg/ml in Soln. A (if it is delivered in another buffer, dialyze against Soln. A). Mix 1 mg β-galactosidase in Soln. A with 0.25 mg SMCC activated antibody (concentration about 1 mg/ml; for activation see Protocol 4.1.3). Shake at RT for 2 h, dialyze or desalt on a Sephadex G-25 column against Soln. B and concentrate to about 2 mg/ml. Mix 9 vol. of the concentrated conjugate with 1 vol. Soln. C and store without further purification at 4 °C.

4.1.4.1 Enzyme Reaction of β-Galactosidase

Solutions/Reagents

A ***substrate buffer***: 3 mM p-nitrophenyl-D-galactoside (M_r 301.3), 10 mM magnesium acetate, 10 mM β-mercaptoethanol (M_r 78,13; ϱ 1.114 g/ml) in TBS, pH 7.5
B ***stop solution***: 0.1 M EDTA in 2 N NaOH
TBS

In a well of a microtiter plate incubate 100 µl of an appropriate dilution of β-galactosidase conjugate at RT for 30 min. Wash thoroughly with TBS and start enzyme reaction by addition of 100 µl Soln. A per well. Stop after 5–20 min (the time period has to be the same for all wells) by addition of 100 µl Soln. B per well. Mix and read O.D. at 405 nm.

4.1.5 Carbodiimide Coupling of Peptides to Carrier Proteins with 1-Ethyl-3-(3-Dimethylaminopropyl)-Carbodiimide (EDAC, EDC)

Solutions/Reagents

A 0.1 M MES (4-morpholinoethanesulfonic acid) pH 4.7

Dissolve 4 mg of carrier protein (KLH, ovalbumin, BSA or the like) in about 400 µl of Soln. A[3]. Add a solution of 4 mg peptide in 400 µl Soln. A and mix well. The molar ratio should be at least 10 moles of peptide per mole carrier protein.

[3] If KLH is used, the buffer should contain 0.9 M NaCl.

Universal coupling protocol

Dissolve about 20 mg EDAC hydrochloride (M_r 191.7) in 200 µl Soln. A immediately before use and mix vigorously with the peptide-carrier solution. Shake at RT for 2 h.

If necessary, remove surplus reagents and/or make buffer exchange by gel filtration on Sephadex G-25 or dialysis. Concentrate the eluate and the dialysate, respectively, to about 1 ml.

As a variant, activate the peptide separately first and couple then to the carrier: Dissolve the peptide to 1 mg/ml in ddH_2O, add 10 mg EDAC hydrochloride per milligram of peptide, and adjust pH to 5.0. Incubate at RT for 5 min and correct the pH with diluted NaOH during this period. Then add the same volume of carrier protein solution. The amount of carrier protein should be in a ratio of 40 moles of COOH groups per mol peptide (ovalbumin: M_r 42.7 kD, 31 Asp, 48 Glu/Mole; BSA: M_r 67.7 kD, 54 Asp, 97 Glu/Mole). Shake at RT for 4 h and stop the reaction by addition of 1/10 volume of 1 M sodium acetate buffer, pH 4.2. Free the sample from surplus reagents by gel filtration or dialysis and concentrate to about 1 ml by ultrafiltration.

Amount and volume of the conjugate are sufficient for immunization of two rabbits (first immunization 200 µl each, first, second, and third boost, 100 µl each).

References

Hermanson G (1996) Bioconjugate techniques. Academic Press, San Diego. p 170

4.1.6 Conjugation of Horseradish Peroxidase (Glycoproteins) by Periodate Oxidation

Glycoproteins, such as horseradish peroxidase, are coupled selectively to other proteins or NH_2-groups bearing molecules via oligosaccharide side chain oxidation. The vicinal OH groups of oligosaccharide residues are oxidized by periodate to aldehyde groups, which react with amines to form imines (Schiff bases).

Solutions/Reagents

A 0.15 M sodium metaperiodate (32 mg/ml ddH_2O)
B 10 mM sodium acetate buffer, pH 4.5
C 0.2 M sodium carbonate buffer, pH 9.5
D 10 mM sodium carbonate buffer, pH 9.5
E 4 mg/ml $NaBH_4$ or $NaBH_3CN$ or 50 mM ascorbic acid, freshly prepared in ddH_2O
F 10 mg/ml BSA in PBS
PBS

Dissolve 2 mg purified enzyme ("Reinheitszahl" RZ $\approx$ 3; RZ = A_{403}/A_{274} at 0.5–1 mg/ml; for purification see Protocol 3.5.2.4) in 0.5 ml ddH_2O. Add 25 µl Soln. A to the stirred enzyme solution.

> ***Important!*** *A color change has to occur. If no change is observed, discard the HRP.*

Continue stirring at RT for 20 min and dialyze twice against 100 vol. Soln. B each for 3 h. Transfer the dialysate into a fresh reaction tube, add quickly 10 µl Soln. C, vortex, and add immediately 5 mg antibody in Soln. D (about 10 mg/ml).

Incubate at RT on a shaker for 2 h, then add 1/10 vol. of Soln. E and mix again. Allow the reduction of SCHIFF bases at 4 °C for 2 h, and then dialyze against PBS overnight. Instead of dialysis, a gel filtration on a Sepahdex G-25 column, equilibrated with PBS, is possible. Concentrate the brownish dialysate and eluate, respectively, and add BSA to a final concentration or 10 mg/ml, mix with an equal volume of glycerol, and store at 20 °C.

> ***Important!*** *Never bring HRP solutions in contact with sodium azide!*

Instead of dialysis or gel filtration, an affinity chromatography on Concanavalin A Sepharose (cf. Protocol 3.6.2.4) is recommended, because on the one hand, no conjugated antibody is removed, and on the other hand, the sugar used for elution stabilizes the enzyme-antibody conjugate in solution.

Take the enzymatic reaction as described in Protocols 2.5.4.1 and 4.13.1.

References

Nakana PK (1980) In: Nakamura RM, Dito W, Trucker ES III (eds.) Immunoassays. Liss, New York, p 157

Huson L, Hay F (1989) Practical Immunology, 3rd ed. Blackwell, Oxford, p 44

4.1.7 Conjugation of Peptides to Carrier Proteins Using Glutaraldehyde (Two-Step Procedure)

Solutions/Reagents

A PBS

B 50% glutaraldehyde (w/v) in pure water (stabilized solution)

Universal coupling protocol

Dissolve 5 mg of ovalbumin, BSA, or another suited carrier protein in 100 µl ddH_2O. The pH has to be 6, because aggregation occurs above pH 7. Add Soln. B to a final concentration of 2.5% glutaraldehyde and stir at RT for 30 min. Filtrate the reaction mixture on a PD-10 or Sephadex G-25 column, equilibrated with ddH_2O, and collect the activated protein in the void volume[4].

[4] A stock of glutaraldehyde-activated BSA with a concentration of 10 mg/ml may be stored frozen in aliquots. Activated BSA of high stability with about 20 moles aldehyde per mole BSA is made by a dialysis procedure given by ZEGERS et al. (1990) J Immunol Meth 130:195.

Dissolve the required amount of peptide (1 mole peptide per about 50 moles of lysine residues of the carrier, e.g., 60 μmol of peptide per 5 mg ovalbumin) in PBS and mix with the activated carrier. Stir at RT for 1 h and block with 10 mg/ml of solid $NaBH_4$ at RT for 20 min. Alternatively, reduce (block) the formed azomethines (Schiff bases) to secondary amines by addition of ascorbic acid (final concentration 5 mM).

Dialyze or desalt by gel filtration, aliquot, and freeze until immunization.

References

Kerr MA, Thorpe R (eds.) (1994) Immunochemistry Labfax. Academic Press, Oxford. p 66

4.1.8 Conjugation of HRP to Antibodies with Glutaraldehyde

Another possibility of conjugating HRP to proteins is the glutaraldehyde-mediated coupling. We prefer the oxidative coupling (see Protocol 4.1.6) because it does not influence the polypeptide backbone of the enzyme, but glutaraldehyde conjugation gives satisfying results, too.

Solutions/Reagents

A 25% glutaraldehyde (w/v) in ddH_2O
B 1 M sodium carbonate buffer, pH 9.5
C 0.2 M lysine in pure water
PBS

Dissolve 10 mg of purified HRP ("Reinheitszahl" RZ $\approx$ 3) in 0.2 ml PBS, then add 10 μl Soln. A. Incubate at RT overnight, dialyze against 25 ml PBS and fill up to 1 ml with PBS.

Mix 1 ml of the antibody solution (5 mg/ml PBS) with the activated HRP and with 0.1 ml Soln. B. Allow to react at 4 °C overnight, then add 0.1 ml Soln. C and incubate at RT for 2 h. Finally, dialyze against 25–30 ml PBS.

Purify the conjugate by affinity chromatography on Concanavalin A Sepharose (Protocol 3.6.2.4), add glycerol to 50% final concentration and store aliquots at −20 °C. The aliquots are stable for months.

References

Dent AH, Aslam M (1999) The preparation of protein-protein conjugates. In: Aslam M, Dent AH (eds.) Bioconjugation. Protein coupling techniques for the biomedical sciences. Macmillan Reference Ltd., London, p 216
Boorsma DM (1983) In: Cuello AC (ed.) Immunhistochemistry. IBRO Handbook Series: Methods in the neurosciences, vol. 3. Wiley, Chichester, p 87

4.1.9 Alkaline Phosphatase-Immunoglobulin Conjugate (Glutaraldehyde Protocol)

The advantage of the use of alkaline phosphatase (AP) lies in the high purity of commercially available enzyme. But it is difficult to get such highly specific activities as for HRP, because there is no simple procedure for separation of unbound antibodies from the conjugate.

A 1% glutaraldehyde (w/v) in ddH_2O
B 50 mM Tris·HCl, pH 8.0, 1 mM $MgCl_2$, 0.02% NaN_3(w/v)
PBS

Mix 5 mg of affinity-purified immunoglobulin in 1 ml PBS with 10 mg AP[5] in 1 ml PBS. The molar ratio should be 1 mol IgG: 2 mol AP (M_r IgG 150 kD, M_r AP 140 kD). Dialyze the mixture twice at 4 °C against 100 ml PBS for 6–9 h each.

Add 0.15 ml Soln. A to the dialysate. Agitate at RT for 2 h and continue at 4 °C overnight. Dialyze three times against 100 ml PBS for 3 h each. Bring the dialysate up to 2 mg BSA and fill up to 10 ml with Soln. B. Aliquote and store at −20 °C.

A separation of conjugated AP from unbound enzyme in EIA is not necessary, because free AP is removed during the washing steps.

4.1.9.1 Enzymatic Reaction of Alkaline Phosphatase from Calf Intestine

This protocol gives soluble colored reaction products used for enzyme immunoassays in test tubes. A procedure yielding insoluble reaction products is given in Protocol 2.5.4.2.

Solutions/Reagents

A 2.7 mM p-nitrophenyl phosphate (pNP; M_r 371.15, disodium salt hexahydrate), 0.5 mM $MgCl_2$, 0.1 M diethanolamine, pH 9.5, in ddH_2O. Use highly pure colorless pNP
B 0.1 M EDTA in 2 N NaOH

Warm 0.5 ml of Soln. A up to 37 °C and add 5 µl conjugate and conjugate dilution in PBS, resp. Stop with 100 µl Soln. B after 5 to 10 min. If the assay is done in parallels, each probe has to be stopped after exactly the same time. Read the O.D. at 405 nm.

4.1.10 Labeling of Immunoglobulins with Fluorescent Dyes

Solutions/Reagents

A 0.2 M carbonate buffer, 0.5 M NaCl (2.12 g anhydrous Na_2CO_3, 2.92 g NaCl in 100 ml ddH_2O), adjust to pH 9.2 with HCl
PBS

[5] If AP is supplied as ammonium sulfate suspension, spin an aliquot containing 10 mg with 12 000 × g at 4 °C for 15 min. Discard the supernatant completely, add 1 ml PBS, followed by 1 ml of antibody solution in PBS.

Add 33 µl of fluorescein isothiocyanate (FITC; tetramethylrhodamine isothiocyanate, TRITC, or another fluorescent dye isothiocyanate derivative is used the same way), 50 mg/ml in DMF, to 1 ml of 5 mg/ml IgG in Soln. A. Shake at RT protected from light for 1 h. Remove surplus FITC and its hydrolysis products on a Sephadex G-25 column, equilibrated with PBS. The conjugate appears in the void volume and should be concentrated by ultrafiltration. Add sodium azide to a final concentration of 0.02% (w/v) and glycerol up to 10% (w/v) and store at 4 °C.

A selection of fluorescent dyes is given in Table 4.3.

Calculate the degree of substitution F/P (nmoles FITC per nmoles IgG) using the following equation:

$$F/P = \frac{A_{495} \cdot V \cdot 13.1515 \cdot 160}{P}$$

A_{495}, absorption of the sample at 495 nm and 1 cm path; V, total volume of the conjugate in milliliters; P, total amount of IgG in micrograms.

A good conjugate has an A_{495}/A_{280} quotient of 0.3 – 1.0.

The protein concentration P is determined by measuring the sample dissolved in PBS (reference: PBS) at 280 nm with consideration to the fluorochrome-specific absorption at 495 nm:

$$\text{Fluorescein:} \quad P = \frac{A_{280} \cdot (0.35 \cdot A_{495}) \cdot \text{dilution}}{1.38} \ (\text{mg/ml})$$

$$\text{Tetramethylrhodamine:} \quad P = \frac{A_{280} \cdot (0.56 \cdot A_{550}) \cdot \text{dilution}}{1.38} \ (\text{mg/ml})$$

The ratio F/P (moles fluorochrome per milligram of protein) is calculated by

$$\text{Fluorescein} \quad F/P = \frac{2.87 \cdot A_{495}}{A_{280} - (0.35 \cdot A_{495})}$$

$$\text{Tetramethylrhodamine} \quad F/P = \frac{4.47 \cdot A_{550}}{A_{280} - (0.56 \cdot E_{550})}$$

P: protein concentration in mg/ml; A_{280}, A_{495}, A_{550}: absorption at 280, 495 (fluorescein), and 550 nm (tetramethylrhodamine), respectively; 1-cm path, blank PBS; dilution: dilution of the sample (dilution is 101, if 10 µl of conjugate solution are added to 1000 µl).

References

Friemel H (ed.) (1991) Immunologische Arbeitsmethoden. Gustav Fischer, Jena, p 530

Hermanson GT (1996) Bioconjugate Techniques. Academic Press, San Diego, p 303

Hudson L, Hay FC (1989) Practical Immunology. 3rd ed. Blackwell Sci. Publ., Oxford, p 34

Table 4.3. Fluorescent dyes[6]

Fluorescent dye and derivative	Excitation (nm)	Emmis-sion (nm)	Coupling to
2′,4′,5′,7′-Tetrabromosulfon fluorescein isothiocyanate	528	544	NH_2
2′,7′-Dichlorofluorescein isothiocyanate	510	532	NH_2
4′,5′-Dichloro-2′,7′-dimethoxy fluorescein	522	550	(DNA)
4′,6-(Diimidazoline-2-yl)-2-phenylindole (DIPI)	364	455	
Acridin Orange	460	650	(RNA)
Acridin Orange	500	526	(DNA)
Alexa 350 succinimidyl	346	442	NH_2
Alexa 430 succinimidyl	431	541	NH_2
Alexa 488 succinimidyl and maleimide	495	519	NH_2 and SH
Alexa 532 succinimidyl and maleimide	531	554	NH_2 and SH
Alexa 546 succinimidyl	556	573	NH_2
Alexa 568 succinimidyl	578	603	NH_2
Alexa 594 succinimidyl and maleimide	590	617	NH_2 and SH
BODIPY	500	506	NH_2
BODIPY FL	505	513	NH_2
BODIPY FL Br_2	533	548	NH_2
Cy2, NHS ester	489	506	NH_2
Cy3, NHS ester	550	570	NH_2
Cy3.5, NHS ester	581	596	NH_2
Cy5, NHS ester	649	70	NH_2
Cy5.5, NHS ester	675	694	NH_2
Cy7, NHS ester	743	767	NH_2
DY-495-X5-01, NHS ester	493	521	NH_2
DY-505-X5-02, amino derivative	505	530	COOH
DY-633-03, maleimide	637	657	SH
Eosin	535	542	NH_2
Erythrosin	530	555	NH_2
Ethidium bromide	518	605	(DNA)
Fluorescein isothiocyanate (FITC)	494	518	NH_2
FluorX	494	520	NH_2
Oregon Green 488	496	524	NH_2
Oregon Green 514	511	530	NH_2
R-Phycoerythrin	488	578	
PicoGreen	502	523	DNA
Propidium bromide	535	617	DNA
Rhodamin 6G	525	555	NH_2
Tetramethylrhodamine isothiocyanate (TRITC)	555	580	NH_2
Texas red	595	615	NH_2

4.1.11 Protein-Colloidal Gold Conjugates

Protein-labeled colloidal gold probes suit as well as for immunohistochemistry in electron microscopy as for detection of antigens or glycoproteins on blots. There are several protocols for enhancement of electron density and visible color resulting in higher sensitivity.

The size of the colloidal particles depends on reagents and their concentration during reduction of gold salts. Larger particles with size > 15 nm are obtained when citrate was the reduction agent, smaller particles result from tannic acid reduction. The unlabeled gold sol is stable in ddH_2O, but traces of salt let collapse the colloid. Covering with proteins increases stability, but especially colloids of small diameter tend to aggregate irreversibly during time; therefore, it is recommended to prepare sufficient amounts of colloidal gold and to store it unlabeled, and the protein-coated gold should be made as fresh as possible.

4.1.11.1 Preparation of Colloidal Gold Sol

Solutions/Reagents

- A ddH_2O, filtered with 0.45 nm membrane filter and stored dust free and salt free in plastic bottles
- B 1% gold(III)chloride (tetrachloroauric acid) (w/v) in Reagent A
- C 1% trisodium citrate (w/v) in Reagent A
- D aqua regia (chloronitrous acid): 1 vol. concentrated HCl+3 vol. concentrated HNO_3
- E 1% tannic acid (w/v) in Reagent A; clear by centrifugation and store in a refrigerator for 1–2 days
- F 1% K_2CO_3 (w/v) in Reagent A

To get gold colloids with reproducible size, the given volumes have to be observed exactly. Store the very hygroscopic gold(III)chloride in a desiccator on anhydrous calcium carbonate powder.

Clean an Erlenmeyer flask and a Teflon-coated stirrer bar with Soln. D and wash thoroughly with Reagent A.

Pour 198 ml Reagent A into the flask, add 2.0 ml Soln. B and heat by strong stirring to boiling. Add quickly 2.0 ml Soln. C and reflux by strong stirring for 10 min. The color of the solution changes from pale yellow via gray to purple. The absorption maximum of the prepared colloid with a mean diameter of 20 nm is at 530 nm. Formation of aggregates is characterized by a shift to larger wavelength.

The cooled colloid is filled into and stored in a plastic flask cleaned with Soln. D.

[6] A lot of dye names are trademarks. For detailed information see, for example, Molecular Probes (www.probes.com), BD Biosciences (www.bdbiosciences.com/spectra/), Dynamics GmbH (www.dynomics.de), or GE Amersham Biosciences (www.amershambiosciences.com/life).

References

Frens G (1973) Nature (Phys Sci) 241:20

Gold colloid with smaller diameter is made by reduction of the tetrachloroauric acid with a citrate/tannic acid reaction mixture given in Table 4.4.

Mix 79 ml Reagent A with 1 ml Soln. B and heat to 60 °C while stirring. Prepare the reduction mixture according to Table 4.4 and heat to 60 °C, too. Add quickly the hot reaction mixture to the gold solution and continue heating and stirring for 15 min. When the color has changed to red, reflux for 10 min.

Table 4.4. Reduction mixtures for preparation of colloidal gold (5–10 nm)

Diameter (nm)	Solution A	Solution C	Solution E (ml)	Solution F
≈ 10	16	4	0.09	0
≈ 6	16	4	0.4	0
≈ 4	16	4	2	2

References

Leunissen JLM, De Mey JR (1989) Preparation of gold probes. In: Verkleij AJ, Leunissen JLM (eds.) Immuno-gold labeling in cell biology. CRC Press, Boca Ratan, p 3

4.1.11.2 Adsorption of Protein to Colloidal Gold

Solutions/Reagents

- A 5 mM buffer solution with a pH near to the pI of the respective protein
- B 2 M KCl in pure water
- C 5% Carbowax 20 M (polyethyleneglycol with M_r 20.000)[7]
- D 0.05% Carbowax 20 M (w/v), 0.05% Tween 20 (w/v) in PBS PBS

The adsorption of protein on colloidal gold occurs in a relative small pH range. Immunoglobulins are bound at pH 7.4, Protein A has an optimum at pH 6.5. For adjusting the pH, the gold colloid as well as the protein solution are dialyzed twice at RT against a 100-fold volume of Soln. A for 1 h each.

Determine the amount of protein necessary for saturation of the gold sol: Pipet 0.5 ml of the dialyzed colloidal gold into Eppendorf tubes and add 0.1 ml of a dilution series of the dialyzed protein to

[7] There are differences in polyethylene glycols of different manufacturers. If Carbowax 20 M is not available, check different lots of PEG 200 000.

Determination of optimal dilution

the gold. Shake at RT for 15 min, then add 0.2 ml Soln. B to each tube. The color of the red gold sol switches to blue, if the amount of protein is not sufficient. Read at 650 nm and plot O.D. against amount of protein. The optimal amount is that dilution which gives just no color change.

Dilute the protein solution to a concentration corresponding to the 1.5-fold of the optimal dilution. Pour 1 vol. of the diluted protein into a centrifuge tube and quickly add 5 vol. of dialyzed gold colloid. Vortex and incubate at RT for 15 min. Add 0.1 vol. Soln. C per 10 ml protein gold mixture and vortex again.

Spin with 11 000 × g at RT for 30 min and resuspend the pellet into a volume of Soln. D corresponding to the original volume of the gold sol. Spin again and carefully resuspend the pellet in 1/10 volume of Soln. D.

The red solution is stable at 4 °C for 2-3 weeks. If sediment occurs, spin with 500 × g for 5 min and discard the pellet.

For staining of Western blots use a 1:250 dilution in PBS.

References

Geoghegan WD, Ackerman GA (1977) J Histochem Cytochem 25:1187

Goodman SL, Hodges GM, Livingston DC (1980) Scanning electron microscopy 1980 II:133

4.2 Immunization of Laboratory Animals

Use 50–200 µg of antigen (if conjugates are used: of hapten) dependent on species and antigenicity per injection for immunization. If KLH conjugates are used, there is no need for adjuvants. In all other cases, mix the antigen solution with a commercially available adjuvant (the well-known FREUND's complete and incomplete adjurant should not be of first choice since it often induces severe necrosis). For the ratio of antigen solution to adjuvant, read the instructions for use given by the supplier. If the antigen is prepared by electrophoresis, mince the cutted gel band thoroughly in a small volume of PBS and immunize by intramuscular injection of the suspension into the hind legs nearby the lymph nodes or intradermal injection into the back. To treat animals with care, do not apply more than 150 µl of gel suspension per kilogram body weight. Never use suspensions or emulsions for intravenous injection! Especially when solutions are injected intravenously, filter these solutions through a 0.22-µm sterile filter.

Immunization using PAGE samples

As conjugates, complexes from biotinylated hapten and streptavidin may be used. In that case perform the detection with avidin-biotinylated hapten because avidin and streptavidin do not cross-react.

Mix 300 µl of antigen or hapten-carrier conjugate in PBS or TBS (without sodium azide or Thimerosal and without SDS) with 1 ml of adjuvant based on mineral oil for a single injection dose of one rabbit (one-third of these volumes for injection of mice). Prepare an emulsion by homogenization with a syringe and a thin needle until a stable white cream is obtained. This water-in-oil emulsion is stable for several hours.

An immunization scheme for rabbits is given in Table 4.5. Do not forget to take some milliliters of blood for pre-immun serum immediately before or after the first immunization.

Table 4.5. Protocol of rabbit immunization

Day	Amount of antigen (μg/animal)	Collected blood (ml)	Remarks
0 (Preimmun serum)		5	
1 (First immunization)	50–200		
14 (First boost)	50–100	5	First test
28 (Second boost)	50–100	5	Second test
35 (Third boost)	50–100	5–50	Third test and/or first antiserum
43–44		25–100	Second antiserum

References

Cooper HM, Paterson Y (2000) Preparation of polyclonal antisera. Current protocols in molecular biology. Wiley, Unit 11.12

Green JA, Manson MM (1992) Production of polyclonal antisera. In: Manson MM (ed.) Immunochemical protocols. Methods in molecular biology, vol. 10. Humana Press, Totowa, N.J., p 1

Harlow E, Lane D (1988) Antibodies. A laboratory manual. Cold Spring Harbor Laboratory, p 92

Peters JH, Baumgarten H (eds.) (1992) Monoclonal antibodies. Springer, Berlin, p 39

4.3 Ammonium Sulfate Fractionation of Immunoglobulins

Stir freshly collected blood with a glass rod to bind fibrin and allow clotting first at RT for 1–2 h and then at 4 °C overnight. Pour the serum into a centrifuge tube and centrifuge at 100–300 × g for 10 min. If the serum is not used within the next day, aliquot the serum, freeze rapidly and store below –20 °C.

Solutions/Reagents

A 1 M Tris · HCl, pH 8.0

B saturated ammonium sulfate in ddH_2O (stir about 100 g $(NH_4)_2SO_4$ in 100 ml ddH_2O for several hours, then allow to equilibrate at RT for at least 2 days; make sure to have a sediment of solid ammonium sulfate within the bottle; adjust pH 7.0 with ammonia)

C PBS

D 0.1 M $Na_2B_2O_7$, 0.02% (w/v) NaN_3 or Thimerosal, pH 8.4

E 10% (w/v) PEG 6000 in D

F 5% (w/v) PEG 6000 in D

Spin two 0.7-ml aliquots of (rabbit) serum at 4 °C and 15 000 rpm for 10 min. Join the supernatants and add 1/10 of volume of Soln. A. Determine volume. Add dropwise and slowly the same volume of Soln. B to give a 50% saturation. Shake at RT for 1 h and spin as described above. Carefully remove the supernatant, discard it, and resolve the precipitate with 1 ml ddH_2O. Determine the volume and add 40% of this volume of Soln. B. Allow to precipitate in the refrigerator overnight. Collect the precipitate by centrifugation, dissolve the pellet with ddH_2O, and dialyze against PBS or TBS. Whereas antibody solutions are stable at 4 °C for a longer period, aliquot the dialysate, determine immunoglobulin content by UV reading (equation d), Protocol 1.1.7, (absorption coefficients see Table 4.6), and store the aliquots at −70 °C.

For fractionation with low contamination by serum proteins of sera from other species the percentage of saturation given in Table 4.7 should be used.

Alternatively, immunoglobulins may be purified by affinity chromatography on Protein A, Protein G, Protein L or thiophilic

Table 4.6. Absorption coefficients of immunoglobulins of different species at pH 7.4

Species	Type of IgG	$A_{1cm}^{1mg/ml}$	λ (nm)
Chicken (yolk)	IgY	1.35	275
Goat	IgG	1.3	280
	IgM	1.3	280
Guinea pig	IgG1	1.357	278
Man	IgG	1.38	280
	IgM	1.45	280
Mouse	IgG	1.34	280
	Fab	1.4	278
	IgA	1.35	275
Rabbit	IgG	1.35	280
Rat	IgG	1.46	280
	IgM	1.25	280
Sheep	IgG	1.22	280

Data from Fasman G (ed.) (1992) Practical handbook of biochemistry and molecular biology. CRC Press, Boca Raton, Florida, p 265

Table 4.7. Ammonium sulfate fractionation of sera of different species

Species	Ammonium sulfate saturation (%) 1st Precipitation	2nd Precipitation[a]	3rd Precipitation[a]	Content of Ig (%)
Cat	35	35	30	71
Chicken	35	35	35	73
Goat	45	30	–	83
Guinea pig	40	40	35	74
Hamster	35	35	35	68
Horse	30	30	30	45
Man	50	45	45	
Mouse	40	40	35	75
Pig	35	35	35	72
Rabbit	35	35	35	91
Sheep	35	35	35	84

[a] The amount of ammonium sulfate resulting from the previous precipitation has to be taken into consideration

Data from Jones GL, Hebert GA, Cherry WB (1978) Fluorescent antibody techniques and bacterial applications. H.E.W. Publ., Atlanta

media. Note that loss of immunoglobulin subclasses may occur caused by different binding affinities to the Fc receptors Proteins A and G, and to the light chain receptor Protein L (cf. Table 4.8).

Antibody complexes are precipitated by relatively low concentrations of polyetyleneglycol 6000 (PEG 6000).

PEG precipitation

Dilute serum with Soln. D in a ratio of 1:25 and add the same volume of Soln. E to the dilution. Incubate at 4 °C overnight and spin with 20 000 × g for 20 min. Wash the pellet once with Soln. F and dissolve the precipitate after a further centrifugation in PBS.

References

Harlow E, Lane D (1988) Antibodies: a laboratory manual. Cold Spring Harbor Laboratory, New York, p 298

Digeon M, Laver M, Riza J, Bach JF (1977) J Immunol Methods 16:165

4.4 Removal of Unspecific Immunoreactivities

In Western blots and in immunofluorescence, antisera often give unspecific reactions, especially when rabbit sera are used. These unspecifities may be suppressed by pretreatment with liver powder.

Add about 50 mg of liver powder to 1 ml of antiserum and shake at RT for 15 min. Then spin at 10 000 × g. The supernatant is ready for preparing working dilutions.

Table 4.8. Binding affinities of immunoglobulin subclasses from different species to immobilized Protein A, Protein G, Protein L, and thiophilic chromatography supports

Species	Antibody class	Affinity to Protein A	Protein G	Protein L	Thiophilic medium
Chicken	IgY	–	–	–	++
Cow	IgG1	+	+++	–	
Cow	IgG2	+++	+++	–	
Cow	Total IgG	++	+++	–	++
Goat	IgG1	+	+++	–	
Goat	IgG2	+++	+++	–	
Goat	Total IgG	+	+++	–	++
Guinea pig	Total IgG	+++	++		
Horse	IgG(ab)	+	–		
Horse	IgG(c)	+	–		
Horse	IgG(T)	–	+++		
Horse	Total IgG	++	+++		++
Human	Fab	–	–	+++	
Human	IgA	+	–	+++	
Human	IgD	–	–	+++	
Human	IgE	+	–	+++	
Human	IgG1	+++	+++	+++	++
Human	IgG2	+++	+++	+++	++
Human	IgG3	+	+++	+++	++
Human	IgG4	+++	+++	+++	++
Human	IgM	+	–	+++	
Human	ScFv	–	–	+++	
Human	Total IgG	+++	+++		
Mouse	IgG1	+	++	+++	++
Mouse	IgG2a	+++	+++	+++	++
Mouse	IgG2b	+++	+++	+++	++
Mouse	IgG3	++	+++	+++	++
Mouse	IgM	–	–	+++	
Mouse	Total IgG	+++	+++	+++	++
Pig	Total IgG	+++	+	+++	++
Rabbit	Total IgG	+++	+++	+	++
Rat	IgG1	+	++		++
Rat	IgG2a	–	+++		++
Rat	IgG2b	–	+		
Rat	IgG2c	+++	+++		
Rat	Total IgG	+	++	+++	++
Sheep	IgG1	+	+++	–	
Sheep	IgG2	+++	+++	–	
Sheep	Total IgG	+	+++	–	++
–	Recombinant Ab	–	++	++	

– No binding, + weak binding, ++ medium binding, +++ strong binding

4.4.1 Preparation of Tissue Powder (Liver Powder)

Solutions/Reagents

A 10 mM phosphate buffer, pH 6.5
acetone
diethylether

Wash pig liver with water and remove as much blood as possible. Homogenize chopped pig liver in a blendor with the same volume of Soln. A and cool the homogenate to 0 °C. Pour slowly the homogenate to 10 vol. of acetone cooled to −15 °C. Allow to settle at −15 °C and filter the sediment on a Büchner funnel. Wash the tissue twice with three volumes of cold acetone, followed by a wash with cold diethylether. Spread the dehydrated tissue on filter paper, grind up, allow the solvent to evaporate and store it in a desiccator containing anhydrous calcium chloride at 4 °C. The white liver powder remains stable for months.

References

Kleber H-P, Schlee D, Schöpp W (1997) Biochemisches Praktikum. 5th ed., G. Fischer, Jena, p 24

4.5 Preparation of Egg Yolk IgY Fraction

Solutions/Reagents

A TBS: 20 mM Tris·HCl, 0.5 M NaCl, pH 7.5
B 10% (w/v) dextransulfate in A
C 1 M $CaCl_2$
D 0.5 M EDTA, disodium salt, pH 7.5
E saturated ammonium sulfate solution (see Protocol 4.3)
solid ammonium sulfate

Separate yolks of chicken eggs from egg white and discard egg white. Wash the yolks carefully with water to remove adhering egg white. Suspend the yolks in 5 vol. Soln. A by vigorous stirring. Precipitate lipids and lipoproteins by addition of 6 ml Soln. B and 15 ml Soln. C per 100 ml yolk suspension. Stir at RT for 30–60 min and spin at 5000 × g for 10 min. Wash the pellet with a small volume of Soln. A (about 20 ml per yolk) and centrifuge again. Combine the supernatants and filter through a paper filter. Add Soln. D to the clear filtrate to a final concentration of 30 mM EDTA.

Add 24.3 g solid ammonium sulfate per 100 ml EDTA-filtrate (yields a 40% saturation) and stir at 4 °C for 30 min. Collect the precipitate by centrifugation. Wash the pellet once with 30% saturated ammonium sulfate (3 vol. Soln. E + 7 vol. ddH_2O), spin down again and dissolve the precipitated IgY in a small volume of Soln. A (about 1/10 of the original yolk volume). Dialyze against TBS and determine the IgY content spectrophotometrically.

References

Schwarzkopf C, Thiele B (1996) ALTEX 13 Suppl. 16:35
Staak C, Schwarzkopf C, Behn I, Hommel U, Hlinak A, Schade R, Erhard M (2001) In: Schade R, Behn I, Erhard M, Hlinak H, Staak C (eds.) Chicken egg yolk antibodies, production and application – IgY-technology. Springer, Berlin, p 65

4.6 Antibody Fragmentation

Since antibodies of different species and different subclasses are variably cleaved by pepsin and papain, a test run is recommended to check incubation time and cleavage conditions. Take samples of different incubation time, freeze rapidly, and monitor the fragmentation by SDS-PAGE at the end of the experiment. Apply the samples both with and without DDT.

4.6.1 F(ab′)2 Fragments from IgG

Solutions/Reagents

A 100 mM sodium acetate buffer, pH 4.5
B IgG, 5 mg/ml in Soln. A
C 2 M Tris pH 8.8
D 10 mM Tris, 0.15 M NaCl, pH 7.6
pepsin

Test of cleavage conditions

Prepare a series of different concentrations of IgG (1–5 mg/ml) from Soln. A and aliquots of Soln. B. Add 5 µg pepsin per milligram IgG and incubate at 37 °C for 15–24 h. Stop proteolysis by addition of 1/10 volume Soln. C. Check by SDS PAGE for optimal IgG concentration and cleavage time (unreduced F(ab′)2: 110 kD; reduced F(ab′)2: doublet at 25 kD; unreduced Fc: 25 kD; reduced Fc: somewhat below reduced F(ab′)2).

Run the preparative cleavage with up to 10 mg IgG using the optimal conditions found in the previous experiment.

Separate the F(ab′)2 fragments by GPC on Sephadex G-100 from degradation products. The F(ab′)2 fragments elute in the void volume.

References

Harlow E, Lane D (1988) Antibodies: a laboratory manual. Cold Spring Harbor, p 630
Kerr MA, Thorpe R (eds.) (1994) Labfax immunochemistry. Academic Press, Oxoford, p 105

Solutions/Reagents

A 100 mM sodium citrate buffer, pH 3.7
B IgG, 2–10 mg/ml in Soln. A
C 1 mg/ml pepsin in Soln. A
D 2 M Tris base

Alternatively, add 20 µg pepsin (from Soln. C) per milligram IgG to Soln. B. Incubate rabbit IgG at 37 °C for 4 h, add 1/10 volume Soln. D and dialyze against TBS or PBS. Apply the dialysate on a Protein A column and collect the passage containing F(ab′)2 fragments.

The preparation of F(ab′)2 fragments from monoclonal (mouse) IgG is similar, but check for optimal cleavage time and use a Protein-G column.

References

Kürzinger K (1993) Enzymatic and chemical modifications: antibody fragments. In: Masseyeff RF, Albert WH, Staines NA (eds.) Methods of immunological analysis. VCH, Weinheim, p 383

4.6.2 Fab′ Fragments (Rabbit)

Solutions/Reagents

A 550 mM Tris, 5 mM EDTA, pH 8.2
B IgG, 5 mg/ml in A
2-mercaptoethanol
iodoacetamide
PBS

Dialyze F(ab′)2 fragments against Soln. A. Add 2-mercaptoethanol to a final concentration of 0.2 M (15 µl/ml). Incubate at RT for 10 min, cool in an ice bath, and add iodoacetamide to a final concentration of 0.3 M (55.5 mg/ml). Incubate on ice for 1 h and dialyze against PBS or desalt on a Sephadex G-25 column, equilibrated with PBS (Fab′ fragments in the void volume).

4.6.3 Fab Fragments (Rabbit)

Solutions/Reagents

A 100 mM sodium acetate buffer, pH 5.5
B IgG, 5 mg/ml in A
C 1 M cysteine
D 20 mM EDTA
papain

Add 1/20 volume Soln. C, 1/20 volume Soln. D and 10 µg papain/mg IgG to Soln. B. Incubate at 37 °C for 8–12 h, and then add 13.8 mg per milliliter assay volume of iodoacetamide. Allow to react at RT for 30 min, then dialyze against PBS or separate on a Protein-A column (Fab fragments in the passage).

4.7 Heidelberger Curve (Precipitin Curve)

If two molecules having at least two binding sites for each other will interact, you will find a concentration range characterized by forming large aggregates. These aggregates are easy to precipitate.

If an antigen-antibody complex has to be precipitated by a secondary antibody (e.g., antigen bound by mAb precipitated by goat anti-(mouse-IgG) antibody), this range of equivalence must be known. For estimation of the range of equivalence the precipitin assay is used.

Solutions/Reagents

A 20% polyethyleneglycol 6000 (PEG 6000) (w/v) in ddH_2O
B 0.1 N NaOH, 0.1% SDS (w/v) in ddH_2O
PBS

Prepare a series of 1:1 or 1:2 dilutions of the first component, e.g., rabbit immunoglobulin or serum, in PBS. Pipet 0.5 ml of each dilution into a 2 ml tube. Add 0.5 ml of a dilution of the second component, e.g., goat anti-(rabbit IgG) antiserum 1:100 diluted in PBS, to each of the first dilutions. Vortex and incubate at 37 °C for 1 h or at 4 °C overnight. Enhance precipitation by addition of 0.25 ml of Soln. A.

Spin at 8000 × g for 20 min and wash the pellet twice with PBS. If unlabeled components were used, dissolve the pellet with 0.1 ml Soln. B and determine the protein content. If a radioactive labeled compound was involved, count for radioactivity.

Plot the amount of protein and radioactivity, respectively, against dilution. The maximum of the obtained HEIDELBERGER curve indicates the range of equivalence.

References

Heidelberger M, Kendall FE (1935) J Exp Med 62:697

4.8 OUCHTERLONY Double-Radial Immunodiffusion

4.8.1 Purification of Agar

If no agar of improved purity is present, the agar is purified by several dialyses. Suspend 4 g of agar in 100 ml ddH_2O; adjust pH to 7.0 with diluted hydrochloric acid or sodium hydroxide, and heat carefully in a water bath until a clear solution is obtained. Pour the solution onto a plate yielding a 5- to 10-mm-thick layer. After gelation, cut into about 1-cm^2 pieces and dialyze these pieces in a 100-fold volume of ddH_2O for a week. Change the water daily and add to the last portion 0.02% sodium azide. In a closed box, the agar remains stable at 4 °C for month.

4.8.2 Preparation of Slides

Solutions/Reagents

A 1.0–2.0% agar or agarose (w/v), 0.02% NaN_3 in barbital buffer, pH 8.4 (cf. Protocol 4.9)
B acetic acid/ethanol/water 1:5:4 (v/v/v)

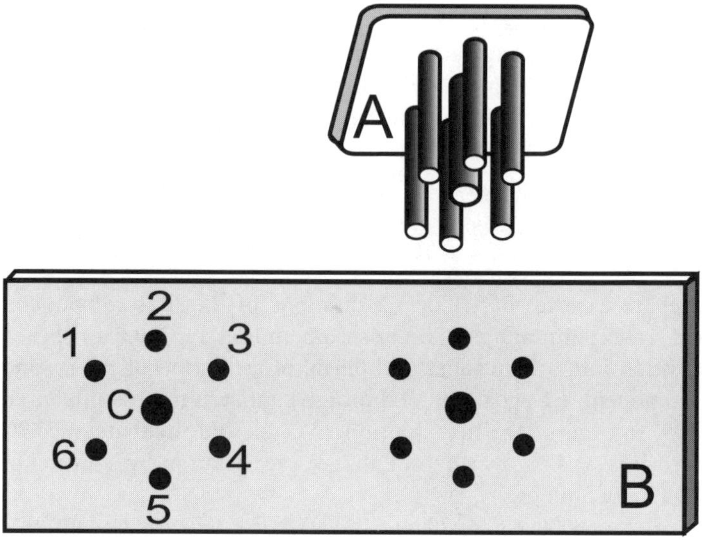

Fig. 4.1. Stamp (*A*) and slide (*B*) with two times of six sample wells (1–6) and larger central well (*C*)

C 0.05% Coomassie Brilliant Blue G250 (w/v) or 0.1% Amido Black 10 B in B
PBS

Melt Reagent A in a water bath or in a microwave oven. Dilute some milliliters of the molten solution with hot water to get 0.5% agar or agarose. Pour this solution onto a carefully defatted slide (3 × 7 cm) and dry in a stream of warm air. Store a stock of these pre-coated slides dustless; they are stable at RT for months.

Pipet 4 ml of the hot molten Soln. A onto the covered slides laying on a horizontally leveled table. After gelation, punch wells using a stamp as shown in Fig. 4.1. Suck out carefully residual material from the wells using a Pasteur pipette connected to a pump.

4.8.3 Immunodiffusion

Depending on the question, fill the central well either with the antigen solution or with antiserum. Place the antigen(s) or antisera or antiserum dilutions into the peripheral wells. If the wells are not filled completely, adjust with PBS.

Place the slides horizontally in a humid chamber, e.g., Petri dishes containing wet filter paper, and store in a refrigerator at least for 24 h, better for 48–72 h.

4.8.4 Visualization of the Precipitin Lines

When diffusion is complete, strong precipitation lines are visible when the slide is viewed obliquely against a black background. Weak lines are visualized by staining.

Incubate the slide three times in PBS for 3 h each. Cover the slide with a sheet of filter paper and dry in air or with a hair dryer. Dampen the slide for a short period and remove the filter paper.

Place the slide into a tray, stain with Soln. C for 5–10 min, and discolor with Soln. B until the background is colorless.

4.9 Immunoprecipitation of Antigens

For analytical purposes, antigens are separated by immunoprecipitation from complex mixtures. So proteins are identified in cell lysates after genome expressions, but also receptors together with their ligands or protein complexes are separated by this method. Mostly, the precipitated immunocomplexes are dissociated and subsequently analyzed by SDS-PAGE. Because it is not possible to get large immune complexes in each case, since for example, antibodies are available directed against only one epitope, antigen-antibody complexes may be precipitated using secondary antibody or antibody receptor (Protein A, Protein G, Protein L) covalently immobilized on chromatographic support.

Solutions/Reagents

A *dilution buffer*: 0.5% Triton X-100 (w/v), 1 mg/ml BSA in TBS

B *lysis buffer*: 0.1 mM PMSF, 0.2 U/ml aprotinin in A
or: 1% Triton X-100 or Nonidet P-40 (NP40), 0.5% sodium deoxycholate (NaDOC), 0.1% SDS (w/v). 150 mM NaCl, 50 mM Tris, pH 7.5

C *sample buffer* (LAEMMLI system): 50 mM Tris · HCl, pH 6.8, 2% SDS (w/v), 10% glycerol (v/v)
precipitation aid: slurry of immobilized secondary antibody or Protein A or Protein G in PBS or TBS

If proteins were expressed within cells of cell cultures, the cells have to be disrupted as gently as possible. For instance, incubate 10^7 to 10^8 cells in 1 ml Soln. B at 0 °C for 30 min. Vortex and spin at 250 × g. Transfer the supernatant into a new tube and centrifuge at 13.000 × g for 10 min.

To reduce unspecific binding, mix a 200-µl aliquot of the clear supernatant with 2 µl pre-immune serum or unspecific antibody and a further 200 µl aliquot with 50 µl precipitation aid. Rock at 0 °C for 1 h and spin at 1000 × g. Transfer the supernatant into a fresh container and fill it up to 1000 µl with Soln. A. Add 0.5–5 µl of the specific antiserum and monoclonal antibody, respectively, and incubate on ice for 1 h. Prepare a second sample containing pre-immune serum instead of antiserum.

Add 50 µl of precipitation aid, 1:1 diluted with Soln. A, and rock at 0 °C for 1 h. Centrifuge at 500 × g for 1 min, discard the supernatant, and wash the pellet four times with 1 ml ice-cold Soln. A each.

Add 20–50 µl Soln. C to the pellet obtained after the last centrifugation. Heat to 95 °C for 5 min (if necessary, add 2-mercaptoethanol or DTT to final concentration of 5% and 10 mM, respectively). Centrifuge and apply the supernatant directly onto the SDS-PAGE gel. The immunoprecipitated protein is identified by staining, Western blot, and/or autoradiography.

References

Åkerström B, Björck L (1989) J Biol Chem 264:19740
Harlow E, Lane D (1988) Antibodies: a laboratory manual. Cold Spring Harbor Laboratory, p 447

4.10 Immunoelectrophoresis

Electrophoresis in agarose or agar gels is, in contrast to SDS-PAGE, a nondenaturing method; therefore, the proteins move according to their net charge, which is at the given alkaline pH mostly negative, and are separated by electrophoretic mobility and not by molecular size. The electropherograms obtained after immunoelectrophoresis are not comparable to SDS-PAGE pictures.

Solutions/Reagents

A 15.4 g sodium diethylbarbiturate (barbital sodium, barbitone sodium, Veronal sodium), 2.76 g diethylbarbituric acid (barbital), 1 g NaN_3 dissolved in ddH_2O, pH adjusted to 8.4 and filled up to 1000 ml

B 1.0–1.5% agar (w/v) in 1:1 diluted Soln. A. Melt before use in a microwave oven or a boiling water bath

Pour 4 ml of molten Soln. B onto pre-coated slides (cf. Protocol 4.8.2). Punch wells according to Fig. 4.2 after gelation, and remove residues of agar from the wells.

Mix the probe 1:1 with Soln. A and add a trace of bromophenol blue. Fill the wells completely with sample.

Place the slide on the cooling plate of a horizontal electrophoresis apparatus, fill the tanks with 1:1 diluted Soln. B, and connect the ends of the slide to the buffer tanks by wetted filter paper wicks. Run the electrophoresis at 6 V/cm for about 45 min. Stop electrophoresis when bromophenol blue is about 4 mm from the the wicks.

Cut the antiserum slots "Ab" after finishing the electrophoresis and remove the agar within the slots. Pour antiserum into the slot(s) and store the slide in a humid chamber at 37 °C for 4–6 h or at 4 °C for 24–48 h. Visualize the precipitation lines as described in Protocol 4.8.4.

References

Hudson L, Hay FC (1989) Practical immunology, 3rd ed. Blackwell, London, p 236
Milford-Ward A (1977) In: Thompson RA (ed.) Techniques in clinical immunology. Blackwell, Oxford, p 1

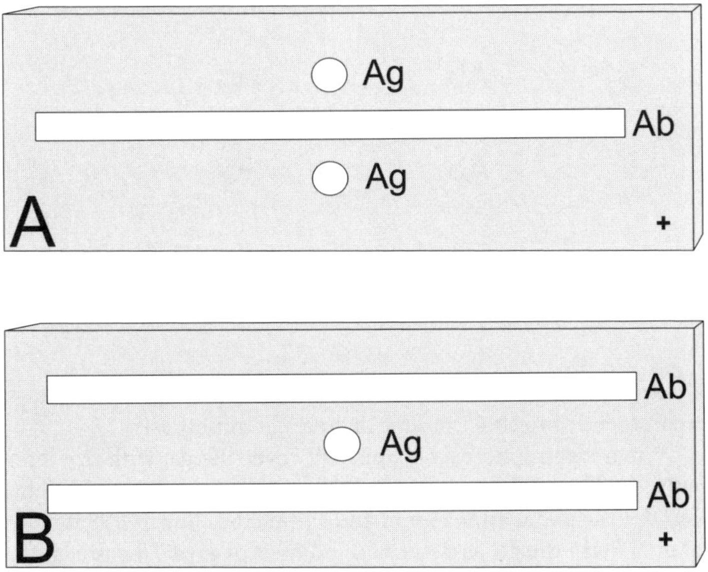

Fig. 4.2. Pattern of immunoelectrophoresis. **A** Two antigen (*Ag*) samples, one antiserum (*Ab*). **B** One antigen, two antisera. + Anode

4.11 Counterelectrophoresis

Solutions/Reagents

A barbital-acetate buffer, pH 8.2, I = 0, 1[8]
B 0.5% agarose (w/v) in ddH_2O
C 1.0% agarose with high endoosmosis (high EEO) (w/v) in 1:1 diluted Soln. A. Melt before use in a microwave oven or a boiling water bath

A synonym for "counterelectrophoresis" is "crossing over electrophoresis."

Pre-coat slides with Soln. B (see Protocol 4.8.2) and prepare the gel by pouring 0.2 ml/cm^2 molten Soln. C onto the slide placed on a leveled surface. Be sure to cover the whole slide with a uniform layer of agarose.

Punch at least two wells of 2- to 3-mm diameter using a Pasteur pipet or a hypodermic needle. The wells are located in the middle of the slide and should have a distance of 5 mm (Fig. 4.3).

Place the slide into a horizontal electrophoresis apparatus, fill the tanks with Soln. A, and connect the small ends of the slide to the electrode tanks by moistened filter paper wicks. Fill the well directed towards the anode with antiserum dilution and pipet antigen solution containing traces of bromophenol blue in that well which is nearby the cathode.

[8] Ionic strength $I = \frac{1}{2} \cdot \sum_{n=1} (c_i \cdot z_i^2)$; c_i, concentration of ion i, z_i, charge of ion i

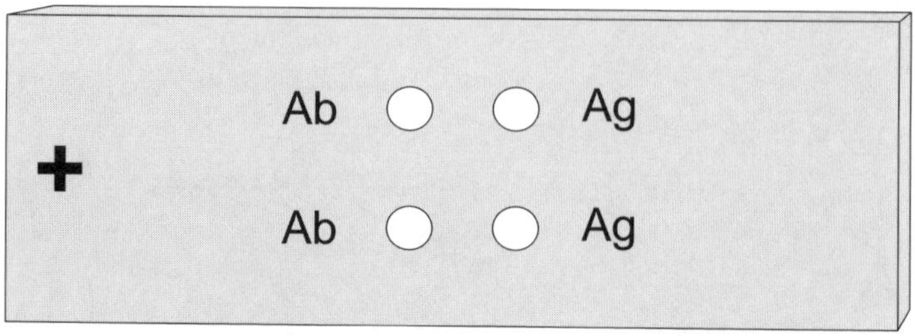

Fig. 4.3. Pattern of counterelectrophoresis. *Ag* antigen well, *Ab* antibody-containing well. + Anode

Run electrophoresis with 20–40 V/cm for about 45 min. It is recommended to cool the slide during electrophoresis.

When electrophoresis is finished, cover the gel with dry filter paper and several layers of tissue paper and press by a weight. Remove the paper after 15 min and agitate the slide in PBS for 2 × 30 min. Stain the gel as described in Protocol 4.8.4. The precipitin bands appear as fine blue lines.

References

Hudson L, Hay FC (1989) Practical immunology, 3rd ed. Blackwell, London, p 241

4.12 Dot-Blot Assay

Mostly plastics, such as nitrocellulose, polyvinylidene fluoride (PVDF), or polystyrene (used for microtiter plates), are used for immobilization of antigens. Because these materials interact with proteins by different mechanisms, sometimes it occurs that the binding material induces or removes epitopes necessary for binding of specific antibodies. So, if no reaction between antigen and antibody is observed, repeat the experiment by using another support (e.g., Nunc offers several types of polystyrene microtiter plates with different binding properties) or bind the antigen via a spacer (e.g., by biotinylation of antigen and immobilization of streptavidin on the support). A dubious positive result should be checked, if possible, by a competition experiment of incubation of the antibody-containing solution in the presence of dissolved (free) antigen.

Test of specifity by competition

Solutions/Reagents

A 0.1% gelatin (w/v) or 5% heat-inactivated calf serum or 0.1% serum albumin (w/v) or 0.2% non-fat milk powder (w/v) or 0.1% Tween 20 (w/v) in PBS

B secondary antibody-HRP conjugate dilution in PBS (dilution for instance 1:1000 to 1:50 000)

C 50 mg/ml 3,3′,4,4′-tetraaminodiphenylether or 3,3′-diaminobenzidine (DAB)or 4-chloro-1-naphthol in DMF or precipitating TMB according to Protocol 2.5.4.1. Stock solutions are stable at RT

> ***Important!*** *If possible, do not use DAB because it is potentially carcinogenic*

D 0.4 ml C, 20 µl 30% H_2O_2 (alternatively: 6 ml 0.1% H_2O_2-urea adduct in ddH_2O; stable for several day at 4 °C), 0.1 ml 10% $CuSO_4$ (w/v), 0.05 ml 10% $NiSO_4$ or $NiCl_2$ (w/v) in 100 ml PBS. Prepare freshly before use
PBS

Indicate the dots with a pencil in a distance of 3–5 mm on a nitrocellulose sheet. Do not touch the nitrocellulose with unprotected fingers! Place the nitrocellulose onto a sheet of filter paper.

Apply 0.1–1 µl of antigen solution, diluted in PBS, on the marks. The diameter of the resulting dot should be not more than 3 mm. Dry at air for 10 min and block the nitrocellulose strip in Soln. A at RT for 15 min.

Incubate the wet strip, or parts of it, in antibody dilution in Soln. A at RT for 30 min. The incubation volume should be 0.5–1 ml per cm^2. This volume may be reduced if the incubation is done in a lockable tube on a roller desk. Wash the strip in a sufficient volume of PBS, at least three times for 5 min each.

Agitate the wet nitrocellulose in 0.5 ml/cm^2 of Soln. B at RT for 15 min and wash three times with PBS. Place the strip in a fresh container and incubate with Soln. D until the color appears. Stop the enzymatic reaction by discarding Soln. D and washing with water.

For a semiquantitative assay, cut each dot and put it into separate test tubes. Perform the enzymatic reaction with TMB substrate, stop with sulfuric acid and read O.D. at 450 nm. (This procedure is similar to an EIA on microtiter plates.)

4.13 Enzyme Immunosorbent Assay (EIA, ELISA)

The following protocols are not optimized procedures for EIA, but they are suitable for screening, e.g., for antibody titers of sera or mAb cell culture supernatants. A high-performance EIA has to be evaluated with respect to selection of type of microtiter plates, coating concentration, coating conditions, analyte dilution, sample buffer, washing buffer, incubation times and temperatures, conjugate dilution, and substrate composition.

The EIA described in this protocol is a so-called indirect EIA, because the antigen is immobilized onto the surface of the microtiter plate and a second species-specific antibody enzyme conjugate detects the bound antibody of the antiserum. As an example, the enzyme label in the presented protocol is horseradish peroxidase (HRP), but when alkaline phosphatase or β-galactosidase or biotinylated secondary antibodies and streptavidin-enzyme conjugates are used, proceed analogously; only substrate solutions and stop solutions for AP and β-galactosidase have to be different.

4.13.1 Indirect EIA with HRP Conjugate

A 15 mM Na_2CO_3, 35 mM $NaHCO_3$, 0.02% NaN_3 (w/v), 0.001% phenol red, pH 9.6, in ddH_2O

B 0.05% Tween 20 (w/v) in PBS

C 0.05% Tween 20 (w/v), 0.1% serum albumin or gelatin (w/v), 0.001% phenol red in PBS

D 0.1 M citric acid (1.92 g anhydrous citric acid in 100 ml ddH_2O), adjusted to pH 5.0 with NaOH

E 0.40 mg/ml o-phenylenediamin (OPD), 0:5 µl/ml 30% H_2O_2 in D (prepare freshly)

E′ 0.55 mg/ml 2,2′-azino-bis(3-ethylbenzthiazoline-6-sulfonic acid), diammonium salt (ABTS), 0.5 µl/ml 30% H_2O_2 in D (prepare freshly each)

E″ 5 vol 0.1 M sodium acetate buffer, pH 5.0, 1 vol. of 1.2 mg/ml N,N,N′,N′-tetramethylbenzidine (TMB) in ethanol, and 4 vol. of 0.1% H_2O_2-urea adduct[9] (w/v) in ddH_2O are mixed immediately prior use

F 1 M sulfuric acid

F′ 0.1% NaN_3 in 0.1 M Tris, pH 7.4

PBS

Secondary antibody-enzyme conjugate, e.g., species-specific anti-(heavy chain)-IgG-HRP conjugate

Dissolve the antigen (pure protein or hapten-carrier conjugate) with a concentration between 0.1 and 50 µg/ml in Soln. A. Pipet 0.1 – 0.2 ml of the antigen solution into the wells of a 96-well microtiter plate (high binding capacity) and incubate on a shaker at RT for 1 h or at 4 °C in a humid chamber overnight.

Remove the antigen solution and wash three times with 200 µl Soln. B each. Beat out the last washing solution on tissue paper and block with 200 µl of Soln. C per well on a shaker at RT for 15 min. Remove the blocking solution and wash twice with PBS. Dry the plate at air and store it until use in a sealed plastic bag.

Pipet 100 µl of the antibody or antiserum dilution (for titer determination a serial dilution in PBS), blank (buffer without antiserum) and, if available, controls into the respective wells and incubate on a shaker at RT for 30 min. Remove the antibody dilution and wash three times with 250 µl of Soln. B each.

[9] H_2O_2-urea adduct (M_r 94,07) is used as a stable substance instead of H_2O_2 solution. The respective amounts of sodium percarbonate or sodium perborate are also suitable.

Pipet 100 µl of a suitable dilution of a secondary antibody-HRP conjugate into each well (the dilutions range from 1:500 to 1:100 000, dependent on the quality of the conjugate, and the amount of bound primary antibody; the dilution has to be checked empirically). Incubate on a shaker at RT for 30 min. Wash with Soln. B at least three times to remove traces of unbound conjugate.

Start enzyme reaction by addition of 100 µl Soln. E, Soln. E′, or Soln E″.

> ***Important:*** *The time for enzymatic reaction has to be the same in each well; therefore, pipet substrate solution and stop solution exactly with the same rhythm. Enzyme reactions depend on time and temperature!*

Select the incubation time for an O.D. of the mostly colored well between 0.8 and 2.5. These values are reached usually between 5 and 15 min. Stop color development by addition of Soln. F and Soln F′, respectively. (If OPD or TMB are substrates, stop with Soln. F, in the case of ABTS use Soln. F′).

Read the O.D. at 492 nm (OPD), 450 nm (TMB), and 405 nm (ABTS), respectively, in a plate reader.

References

Avrameas S, Druet P, Masseyeff R, Feldmann G (eds.) (1983) Immunoenzymatic techniques. Elsevier, Amsterdam
Crowther JR (2001) ELISA – Guidebook. Humana Press, Totowa N.J.
Douillard JY, Hoffman T (1983) Meth Enzymol 92:168
Porstmann B, Porstmann T, Nugel E (1981) J Clin Chem Biochem 19:435
Porstmann T, Kiesig ST (1991) J Immunol Meth 150:5

4.13.2 Determination of Enzyme Activity by ELISA

Solutions/Reagents

A 15 mM Na_2CO_3, 35 mM $NaHCO_3$, 0.02% NaN_3 (w/v), 0.001% phenol red, pH 9.6, in ddH_2O
B 0.05% Tween 20 (w/v) in TBS
C 0.1% cold fish gelatin in Soln. B
D 5 vol. 0.1 M sodium acetate buffer, pH 5.0, 1 volume of 1.2 mg/ml N,N,N′,N′-tetramethylbenzidine (TMB) in ethanol and 4 vol. of 0.1% H_2O_2-urea adduct (w/v) in ddH_2O are mixed immediately prior use
E 1 M sulfuric acid
TBS

Dilute anti-(mouse-IgG)-IgG (in the case of monoclonal antibodies; for testing conjugates prepared from antibodies of other origin, use the respective anti-species specific antibody) to 5 µg/ml in Soln. A. Coat the wells of a microtest plate with 100 µl/well of this dilution and incubate at 4 °C overnight. Remove antibody solution and wash once with TBS. Add 150 µl Soln. C per well and incubate at

RT for 30 min. Remove blocking solution, rinse once with Soln. B and add 100 µl/well of a dilution series (e.g., 1:500, 1:1500, 1:4500, 1:13 500, 1:40 500, 1:121 500, 1:364 500 in TBS) of antibody-enzyme conjugate to the wells. Shake at RT for 30 min, remove conjugate solution, knock out the plate on paper tissue, and rinse three times with Soln. B.

Perform the enzymatic reaction with Soln. D for 10–30 min (each well has to react for the same time and at the same temperature) as described above, stop with Soln. E and read O.D. at 450 nm. Plot O.D. against conjugate dilution. The working dilution of the conjugate should give an O.D. of 1.5–2.5.

4.13.3 Isotype Determination by EIA (AP Conjugate)

This protocol describes the use of an alkaline phosphatase conjugate; of course, a HRP conjugate works well, too.

Solutions/Reagents

- A 15 mM Na_2CO_3, 35 mM $NaHCO_3$, 0.02% NaN_3 (w/v), 0.001% phenol red, pH 9.6, in ddH_2O
- B 0.05% Tween 20 (w/v), 0.02% NaN_3 (w/v) in PBS
- C 0.1% serum albumin or gelatin (w/v), in Soln. B
- D 1 mg/ml p-nitrophenyl phosphate (pNP; M_r 371.15, disodium salt hexahydrate), 0.5 mM $MgCl_2$, 0.1 M diethanolamine, pH 9.5, in ddH_2O. Use highly pure colorless pNP and prepare substrate solution freshly
- E 1 N NaOH
- PBS
- antibody recognizing all subclasses of heavy chains of an antibody class of a species, e.g., anti-(mouse-γ-chain)-IgG (goat)
- secondary isotype-specific antibody-AP conjugate, e.g., anti-(mouse-IgG1)-IgG (goat) AP conjugate

Dilute the class-specific capture antibody to 2 µg/ml in Soln. A. Apply 100 µl/well of the coating solution to a microtiter plate and incubate at 4 °C overnight. Remove the liquid from the plate and wash once with Soln. B. Block the wells with 150 µl Soln. C at RT for 2 h. Knock out the blocking solution and wash once with Soln. B. Store the plate in a sealed bag at 4 °C.

Dilute antiserum or hybridoma supernatant in a geometric series with Soln. B. starting at 1:100 and 1:10, respectively. Pipet 100 µl of each dilution in duplicates into the wells and also for positive as well as for negative control of an appropriate dilution of a characterized immunoglubulin. Incubate at RT on a shaker for 1 h. Wash three times with at least 200 µl/well of Soln. B. Add 100 µl of AP conjugate dilution in Soln. B (if the conjugate is not tested and no recommendations are given by the supplier, start with a 1:10 000 dilution). Incubate at RT for 30 min, knock out the conjugate solution, and wash three times with Soln. B. Start enzymatic reaction by addition of 100 µl/well of Soln. D. Stop after 10.0 min at RT with 100 µl/well of Soln. E. Read absorption at 405 nm in a plate reader.

5 Centrifugation

5.1 Speed vs Centrifugal Force Graphs

If particles with densities (specific gravity) larger than the surrounding liquid are distributed in a solution or suspension, and if a force acts which overcomes random thermal motion of these particles, the particles will move to form a stable pellet. The velocity of sedimentation depends on the size of the particle, on the density difference between particle and solution, on the viscosity of the solution, and on the force which causes the movement: the gravitation or the centrifugal force.

Molecules are surrounded by and filled with liquid (mostly water). Because at first view liquids are not compressible, the size of molecules does not change, even at high pressure. But large biologic particles, such as cells or cellular organelles (even proteins), are no bodies uniformly filled with liquid. High pressure produced by high centrifugal forces may irreversibly deform these organelles up to disruption.

The force produced in a rotating body is given as multiples of earth's gravitational field (9.807 m/s^2), and relative centrifugal field (rcf) is calculated by

$$\text{rcf} = \frac{r \cdot \omega^2}{g} = 1119 \cdot r \cdot \left(\frac{\text{rpm}}{1000}\right)^2 \; [\text{x g}]$$

$$\omega = 2 \cdot \pi \cdot \text{rpm}/60$$

r: distance of a particle or part of a centrifuge rotor from the axis of rotation (mm); ω: angular velocity (radians s^{-1}); g: earth's gravitational field (m · s^{-2}); rpm: revolutions per minute (min^{-1})

In a centrifuge rotor g_{max} is the maximal centrifugal force at maximal distance from rotor axis r_{max} (at the bottom of the tube), g_{min} is the minimal force at the top of the tube (r_{min}), and the arithmetic mean of both is called g_{av}. The difference of r_{max} and r_{min} is the sedimentation path length.

In the biochemical lab mostly two types of centrifuge rotors are used. Fixed-angle rotors have tubes with defined angle relative to rotor axis. These angles range from 18 to 45°. With an angel of 0°, i.e., the tube is parallel to the axis, the fixed-angle rotor is a so-called vertical rotor. The other type of rotor is the swinging-bucket rotor.

The tubes of the swinging bucket rotor type are accommodated in a pivoted bucket moving from vertical to horizontal position during acceleration under the influence of the centrifugal field. The sedimentation path length is lowest in vertical rotors and highest in swinging-bucket rotors.

The relations between rotor speed and centrifugal force for different rotors are illustrated by Figs. 5.1–5.3.

Which type of rotor should be used depends on the aim of separation. If insoluble material shall be separated from dissolved compounds, e.g., cell harvesting, pelleting, or clarification of a solution, fixed-angle rotors are the first choice (note that lipid vesicles in aqueous media form a layer on top of the solution during centrifugation). For fractionation of molecules with relatively small differences in size and/or specific gravity, density gradient centrifugation in swinging-bucket or vertical rotors is used, the latter especially in the case of self-generating gradients.

Whereas pelleting does not afford sophisticated technical equipment, density gradient centrifugation should be done in programmable centrifuges with controlled acceleration and deceleration and programmed $\omega \cdot t$ and/or $g \cdot t$. Since most biologic material is more stable at lower temperature, refrigeration is also recommended for high-capacity low-speed centrifuges.

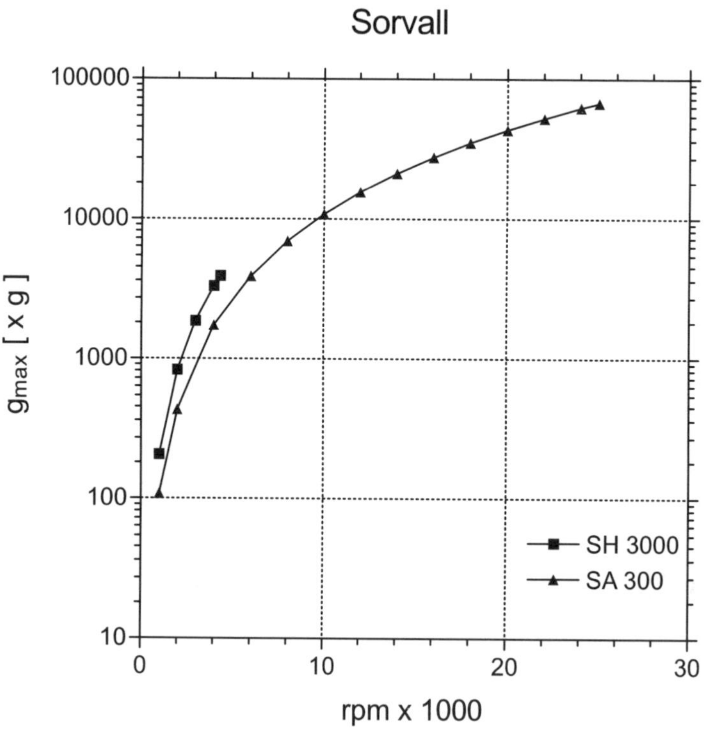

Fig. 5.1. rpm-g graph for fixed-angle (SA 300; 6 × 50 ml) and swinging-bucket (SH 3000; 4 × 750 ml) Sorvall preparative rotor

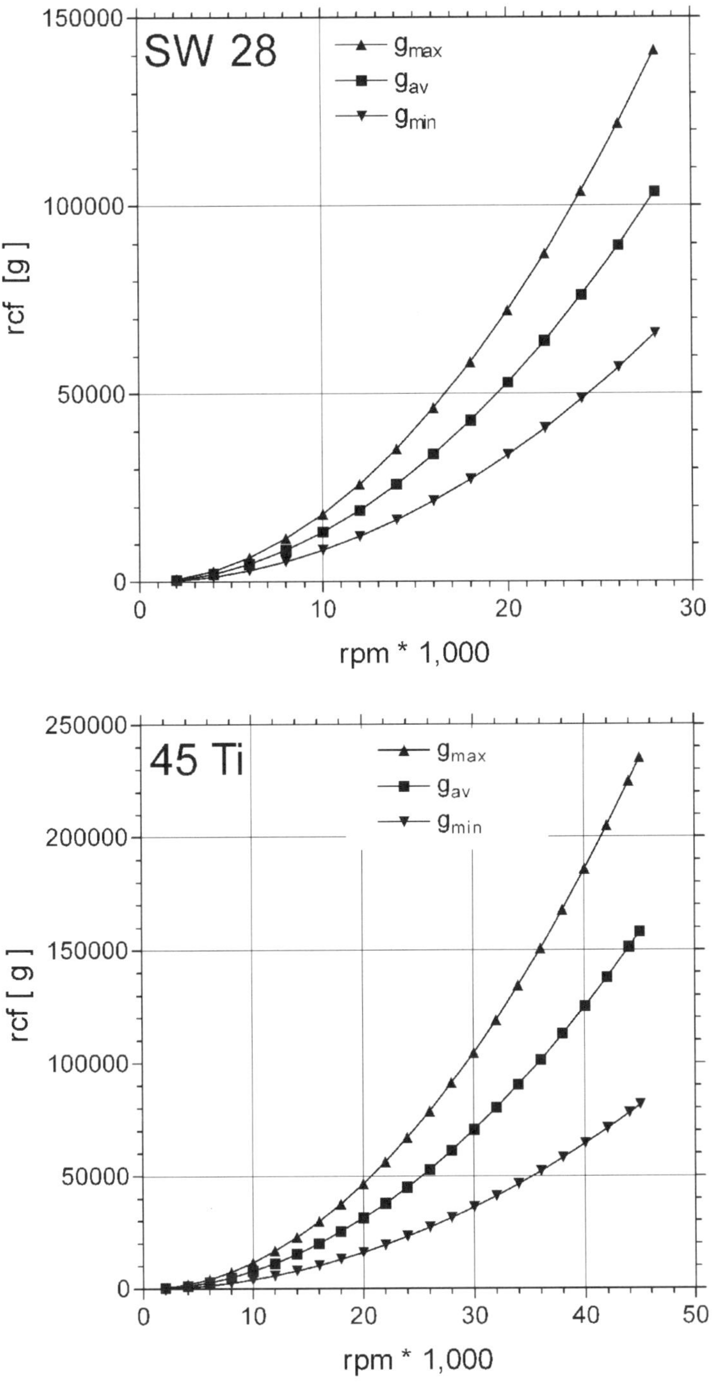

Fig. 5.2. rpm-g graph for fixed-angle (type 45 Ti) and swinging-bucket (SW 28) Beckman Coulter ultracentrifuge rotor

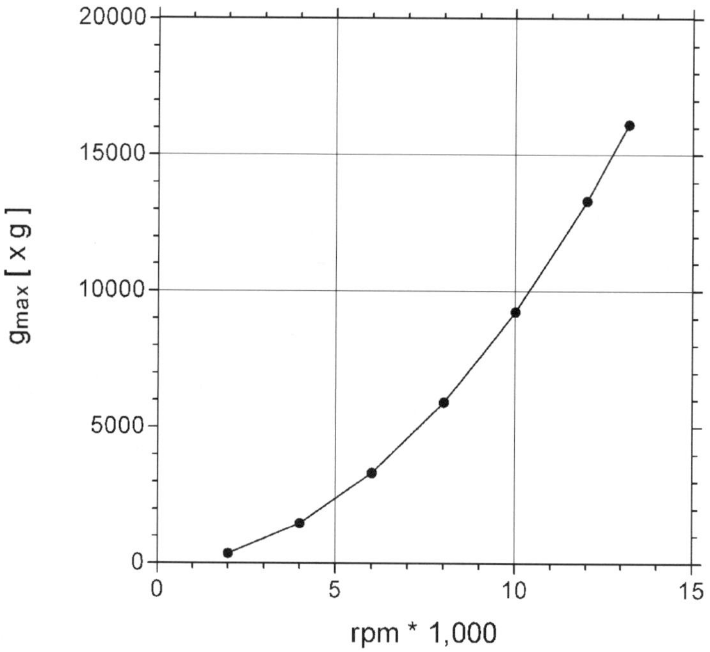

Fig. 5.3. rpm-g graph for fixed-angel Eppendorf tube rotor (Eppendorf 5415 centrifuge; 24 × 1.5 ml)

> ***Important!*** *Rotors, buckets and centrifuge tubes are undergoing enhanced mechanical stress. Handle with care, avoid damage, and check rotors, buckets, bottles, and tubes for any sign of damage prior to running. Balance tubes in diametrically opposed position carefully. The solutions within the tubes should not only have the same volume and weight, but especially in the case of high-speed centrifugation, the same density too. If tubes are placed in threefold geometry, i.e., if a rotor has six positions, for instance, the three filled tubes placed at positions 1, 3, and 5, all tubes must have the same weight and volume. Clean rotors, buckets, and tubes carefully after running.*

References

Osterman LA (1984) Methods in protein and nucleic acid research, vol. 1, Springer, Berlin

Graham J (2001) Biological centrifugation. Bios, Oxford

5.2 Differential Centrifugation

Differential centrifugation is the simplest form of centrifugation: A mixed particle population is subjected to centrifugations with increasing centrifugal forces, resulting in several pellets containing particles with decreasing sedimentation rate.

As an example for differential centrifugation, this protocol gives the preparation of membranes of the sarcoplasmic (intracellular) reticulum (SR) of heart muscle.

Solutions/Reagents

A 10 mM $NaHCO_3$, 5 mM NaN_3, 0.1 mM PMSF[1], pH 7.0
B 0.6 M KCl, 20 mM Tris-maleate, 0.1 mM PMSF, pH 6.8
C 0.25 M sucrose, 5 mM histidine · HCl, pH 6.8

Carry out all operations at 2–8 °C.

Smash deeply frozen pieces of heart (easily done with a hammer, if the tissue is cooled by liquid nitrogen). Thaw portions of 20 g with 100 ml of cold Soln. A and homogenize in a Warring blender twice at maximal speed for 30 s each. Transfer the homogenate into a glass cylinder and homogenize with a Polytron rotating-blades homogenizer at 40% of maximal power, three times for 10 s each. Remove tissue fibers from the homogenizer between each run. Combine the homogenates.

Spin the homogenate in a refrigerated centrifuge at 3000 × g for 20 min. Filter the supernatant through glass wool to remove fat and tissue particles and centrifuge in a pre-cooled fixed-angle rotor at 8700 × g_{max} for 20 min. Repeat filtration and centrifugation.

Centrifuge the combined supernatants with 100 000 × g for 45 min (e.g., Beckman Coulter Type 45 Ti rotor; 35 000 rpm).

Discard the supernatant and resuspend the precipitated SR vesicles in total 1–2 ml of Soln. C, freeze quickly in liquid nitrogen and store aliquots at −70 °C.

References

Harigaya S, Schwartz A (1969) Circ Res 25:781

5.3 Density Gradient Centrifugation

Density gradients may be divided into three types: preformed discontinuous gradients (e.g., Protocol 5.3.1), preformed continuous gradients (e.g., Protocol 5.3.2) or self-generating gradients (e.g., Protocol 5.3.3). Materials used for density gradients are classified as ionic and non-ionic media (Table 5.1).

Gradients made by a distinct substance are not only characterized by the density, but also by viscosity and ionic strength and osmolarity (Table 5.2). Since gradient solutions of high density mostly are concentrated solutions, it should be kept in mind

[1] Add 0.1 M PMSF stock in ethanol, submerging the pipette immediately before use, and mix thoroughly, because PMSF is rapidly inactivated by hydrolysis.

Table 5.1. Substances used in density gradient centrifugation

Ionic	Non-ionic
Cesium chloride	Deuterium oxide ("heavy water")
Cesium sulfate	Sucrose
Cesium trifluoracetate	Glycerol
Potassium iodide	Ficoll
Sodium iodide	Percoll
3,5-Diacetamido-2,4,6-triiodobenzoic acid (metrizoic acid)	2-(3-Acetamido-5-N-methylacetamido-2,4,6-triiodobenzamido)-2-desoxy-D-glucose (Metrizamide)
	Nycodenz
	Iodixanol (OptiPrep)

Table 5.2. Density, viscosity, and concentration of aqueous density gradient solutions at 20 °C

Medium	Density (g/cm^3)	Viscosity ($N \times s \times m^{-2}$)	Conc. (w/w)
Cesium chloride	1.229		25
Sucrose	1.104	2.5	25
Glycerol	1.056	2	24
Metrizamide	1.134	1.9	25[a]
Ficoll	1.09	37	24[a]
Percoll	1.13	10	23

[a] % w/v

that solubility decreases with decreasing temperature, which means that at room temperature, a substance is dissolved, whereas at 4 °C, crystals may be formed which could damage the centrifuge tube at high centrifugal force.

As a consequence of high osmolarity, macromolecules or organelles are dehydrated resulting in altered sedimentation behavior; thus, different buoyant densities for nucleic acids and mitochondria are observed in different density gradient media, as illustrated in Table 5.3. Data of some density gradient media are given in Table 5.4.

5.3.1 Pre-formed Discontinuous Gradient Centrifugation: Isolation of Liver Cell Nuclei

The preparation of cell nuclei from rat liver is an example of the application of a pre-formed stepwise gradient: The tissue homogenate is placed on top of a cushion of concentrated sucrose solution, and only the dense nuclei cross this cushion during centrifugation.

Table 5.3. Buoyant density of mitochondria and DNA from *Micrococcus luteus* (G+C 71%), dependent on the applied density gradient medium

Medium	Mitochondria (g/cm³)	DNA
Metrizamide	1.20–1.25	1.12–1.17
Sucrose	1.19	
Percoll/sucrose	1.09–1.11	
Ficoll	1.14	
Cesium chloride		1.73
Cesium sulfate		1.45
Cesium trifluoracetate		1.63
Potassium iodide		1.51
Sodium iodide		1.55

Solutions/Reagents

A 0.9% NaCl (w/v) in pure water (saline)
B 5 mM $MgCl_2$, 20 mM Tris · HCl, pH 7.4
C 0.25 M sucrose in Soln. B
D 0.34 M sucrose in Soln. B
E 1% Triton X-100 (w/v) in Soln. C

Cool freshly prepared rat liver in ice-cold saline and mince using scissors. Decant the liquid and add 9 vol. (with respect to tissue weight) of ice-cold Soln. B. Homogenize the tissue in a pre-cooled glass-Teflon Potter-Elvhjem homogenizer with about 2000 rpm and eight to ten strokes. Filter the homogenate through several layers of cheesecloth.

Fill lockable centrifuge tubes to 40% with ice-cold Soln. D and cover with the same volume of homogenate. Spin in a refrigerated centrifuge with $1000 \times g_{max}$ at 4 °C for 10 min. Discard the supernatant and suspend the pellet in the original volume of Soln. E. Spin and wash again as described.

Suspend the washed nuclei in 2 ml of ice-cold Soln. C per rat liver.

References

Widnell CC, Tada JR (1964) Biochem J 92:331

5.3.2 Sucrose Gradient Centrifugation: Preparation of Surface Membranes (Sarcolemma, SL) of Heart Muscle Cells

This protocol illustrates the combination of differential and density gradient centrifugation, using a high-capacity low-speed centrifuge as well as an ultracentrifuge.

Table 5.4. Concentration, density, and refractive index of aqueous density gradient solutions (for sucrose see Table 8.17)

Density gradient medium	Conc. (% w/v)	Molarity	Density ϱ (g/ml)	Refract. n_D^{20}
CsCl in water (25 °C)[a,b]	10	0.64	1.079	1.3405
	20	1.39	1.174	1.3498
	30	2.28	1.286	1.3607
	40	3.37	1.42	1.3735
	50	4.70	1.482	1.3885
	60	6.36	1.785	1.4072
	Saturated	7.4	1.91	1.4185
Cs_2SO_4 in water (25 °C)[c]	10	0.3	1.086	1.3438
	20	0.66	1.19	1.3414
	30	1.09	1.317	1.3607
	40	1.62	1.469	1.3718
	50	2.27	1.644	1.3846
	Saturated	3.56	2.01	
Percoll in 0.25 M sucrose	20[d]		1.054	
	40		1.078	
	60		1.102	
	80		1.125	
	Stock solution		1.149	
Percoll in 0.15 M NaCl	20		1.027	
	40		1.051	
	60		1.075	
	80		1.098	
	Stock solution		1.122	
Metrizamide in water (20 °C)[e]	10[f]	0.127	1.025	1.3483
	20	0.253	1.106	1.3646
	30	0.38	1.161	1.3809
	40	0.507	1.216	1.3971
	50	0.633	1.271	1.4133
	60	0.76	1.326	1.4295
	70	0.887	1.381	1.4458
Ficoll in water (20 °C)	10		1.032	1.3469
	15		1.05	1.3545
	20		1.069	1.3625
	25		1.09	1.3713
	30		1.103	1.3801
	35		1.121	1.3893
	40		1.155	1.3986
	50		1.199	1.4171
Nycodenz[g]	0	0	1.030	1.3450
	5	0.061	1.052	1.3512
	10	0.122	1.073	1.3575
	15	0.183	1.095	1.3637

Table 5.4. (continued)

Density gradient medium	Conc. (% w/v)	Molarity	Density ϱ (g/ml)	Refract. n_D^{20}
	20	0.244	1.116	1.3699
	25	0.304	1.138	1.3762
	30	0.365	1.159	1.3824
Nycodenz[h]	0	0	1.005	1.3350
	2.5	0.030	1.018	1.3390
	5.0	0.061	1.031	1.3429
	7.5	0.091	1.044	1.3469
	10.0	0.122	1.056	1.3508
	12.5	0.152	1.069	1.3548
	15.0	0.183	1.082	1.3587
	17.5	0.213	1.095	1.3627
	20.0	0.244	1.108	1.3666
	22.5	0.274	1.120	1.3706
	25.0	0.305	1.133	1.3745
	27.5	0.335	1.146	1.3785
	30.0	0.366	1.159	1.3824

[a] % w/w

[b] $\varrho_{25} = 1.1564 - 10.2219 \cdot n_D^{25} + 7.5806 \cdot (n_D^{25})^2$ for $1.00 < \varrho_{25} < 1.90$ according to Dawson et al. (1986). Data for biological research, 3rd ed., Clarendon Press, Oxford

[c] $\varrho_{25} = 0.9945 + 11.1066 \cdot (n_D^{25} - 1.3325) - 26.4460 \cdot (n_D^{25} - 1.3325)^2$; $\varrho_{25} 1.0047 + 0.28569 \cdot m - 0.017428\, m^2$ for $1.14 < \varrho_{25} < 1.80$; m, molality (moles per 1000 g solution). According to Dawson et al.: loc.cit.

[d] Percentage of stock solution

[e] $\varrho_{20} = 3.350 \cdot n_D^{20} - 3.462$; $\varrho_5 = 3.453 \cdot n_D^{20} - 3.601$ according to: Dawson et al.: loc.cit.

[f] % w/v

[g] In 25 mM sucrose, 10 mM Tris-HCl, pH 7.4; data from Graham J (2001) Biological centrifugation. BIOS, Oxford, p 53

[h] Nycodenz in saline; data from Graham J (2001), loc. cit., p 54

Porcine hearts are obtained from a slaughterhouse. It is absolutely necessary to get the hearts within 5–10 min after killing the animals. The hearts have to be smooth (consistence comparable to liver); firmly contracted muscle should be discarded. Wash the hearts with running cold tap water and remove right ventricles, autricles, atria, fat, large vessels, tendons, and valves. Chop the left ventricles into pieces of about 1.5 cm and freeze immediately in liquid nitrogen. Store the tissue at −70 °C.

Solutions/Reagents

A 0.75 M KCl, 5 mM imidazole, 0.2 mM DTT or DTE, 0.1 mM PMSF[2], pH 6.8

[2] Stock solutions: 0.2 M of DTT and DTE (M_r 154.25) in ddH_2O, 0.1 M PMSF (M_r 174,19) in n-propanol, add immediately before use.

B 10 mM $NaHCO_3$, 5 mM imidazole, 0.2 mM DTT or DTE, 0.1 mM PMSF, pH 6.8
C 0.25 M sucrose, 5 mM imidazole, pH 7.2
D 0.6 M KCl, 10 mM imidazole, pH 7.2
D′ 18% sucrose (w/v) in Soln. D
D″ 27% sucrose (w/v) in Soln. D

All steps have to be done in a cold chamber or in an ice bath.

Crush 250 g of deeply frozen porcine left ventricle and thaw in 900 ml of Soln. A. Homogenize the tissue using a Warring blender (low speed) for 20 s. Spin the homogenate in a Sorval GS-3 rotor or Beckman-Coulter JA-10 rotor at 3000 × g (4500 rpm) for 15 min.

Combine the pellets, suspend in 750 ml of Soln. A and homogenize again for 20 s. Repeat centrifugation. Divide the pellet into 12 portions, fill up each portion with Soln. B to 120 ml, and homogenize each portion using a Polytron homogenizer at 60% of maximal rpm, three times for 15 s each.

Combine the homogenates and spin at 11 000 × g (Sorval GS-3 rotor or Beckman Coulter JA-10 rotor: 8500 rpm) for 20 min.

Centrifuge the combined supernatants at 57 000 × g (Beckman Coulter Type 45Ti rotor: 26 000 rpm) for 40 min; discard the supernatants.

Suspend the pellets (crude membrane fraction) using a glass-Teflon homogenizer in Soln. D to a final volume of 24 ml.

Prepare continuous density gradients from Soln. D′ and Soln. D″, cover with the suspension of the crude membrane suspension and centrifuge with 60 000 × g_{av} and 83 000 × g_{max}, respectively, for 90 min (e.g., Beckman Coulter SW 28 rotor: gradient from 16.5 ml Soln. D′ and 16.5 ml Soln. D″ per tube; 24 000 rpm).

Carefully suck off the white zone located about 1 cm below the meniscus, combine the material from all tubes and dilute 1:2 with Soln. D. Spin at 60 000 × g_{max} for 45 min (e.g., Beckman Coulter Type 50.2Ti rotor: 40 000 rpm).

The resulting pellets are highly purified sarcolemmal membrane vesicles. Suspend the pellets in a small volume of Soln. D, aliquote, and freeze in alcohol-carbon dioxide or liquid nitrogen. Store at −70 °C.

To characterize the preparation, determine marker enzymes, e.g., p-nitrophenyl phosphatase (PNPase), ouabain-sensitive Na,K-dependent ATPase, or dihydropyridine receptor complex (L-type voltage dependent calcium channel).

References

Vetter R, Haase H, Will H (1982) FEBS Lett 148:326

To identify the cellular organelles, enzymes associated with well-established functions are used. The enrichment of an organelle is characterized by an increase of the specific activity (enzyme activity per weight unit of protein). A selection of marker enzymes is given in Table 5.5. It should be noted that these enzymes do not occur in all species and organelles to the same extent, and that their specific activities may differ. Different specialization of parts of a cellular membrane causes uneven distribution of a marker within the membrane, too.

References

Graham JM (1993) The identification of subcellular fractions from mammalian cells. In: Graham J, Higgins J (eds.) Meth Molec Biol, vol. 19. Biomembrane protocols. I. Isolation and analysis. Humana Press, Totowa, N.J., p 1

Table 5.5. Selected marker enzymes for cell fractionation

Cellular component	Marker enzyme	E.C. number
Nucleus	NAD pyrophosphorylase (DNA)	2.7.7.1
Chloroplast	Ribulose-1,5-bisphosphate carboxylase (Chlorophyll)	4.1.1.39
Mitochondrium		
Inner membrane	Cytochrome-c oxidase	1.9.3.1
	Succinic (Fumarate) dehydrogenase	1.3.99.1
Outer membrane	Monoamine oxydase A and B (Cytochrome b5)	1.4.3.4
Matrix	Glutamate dehydrogenase	1.4.1.2
	L-Malate:NAD$^+$ oxidoreductase (m-isoenzyme)	1.1.1.37
Lysosome	Acidic phosphatase	3.1.3.2
	β-Galactosidase	3.2.1.23
	Aryl sulfatase	3.1.6.1
Peroxysome	Catalase	1.11.1.6
Rough endoplasmic reticulum (microsomes)	D-Glucose-6-phosphatase (Ribosome)	3.1.3.9
	NADH-cytochrome c reductase	1.6.99.3
Golgi apparatus	Galactosyl transferase	2.4.1.22
	α-Mannosidase II	3.2.1.114
Plasma membrane	5′-Nucleosidase	3.1.3.5
	Alkaline phosphatase	3.1.3.1
	Alkaline phosphodiesterase I	3.1.4.1
	Oubain-sensitive Na,K-ATPase	3.6.3.9
Cytosol	L-Lactic dehydrogenase	1.1.1.27

Plummer DT (1987) An introduction to practical biochemistry, 3rd ed. McGraw-Hill, London, p 272
Storrie B, Madden EA (1990) Meth Enzymol 182:203

5.3.2.1 Determination of a Marker Enzyme: Ouabain-Sensitive Na,K-ATPase

Solutions/Reagents

- A 0.25 M sucrose, 5 mM histidine, pH 7.2
- B 2% SDS (w/v) in ddH_2O
- C 167.7 mM NaCl, 26.7 mM KCl, 6.3 mM NaN_3, 4 mM $MgCl_2$, 0.13 mM EGTA, 40 mM imidazole. Dissolve required amounts of the solids in about 8 ml ddH_2O, adjust pH 7.2 and fill up with ddH_2O to 10 ml
- ATP 30 mM ATP, disodium salt (18 mg/ml ddH_2O)
- Ou 5 mM ouabain in ddH_2O
- D 4% SDS (w/v), 10 mM EDTA, adjust pH ton 7.5 with NaOH
- E 6 N hydrochloric acid
- F 2.5% ammonium molybdate (w/v) in ddH_2O
- G 10% ascorbic acid (w/v) in ddH_2O
- H mix Soln. E, F, G and ddH_2O in a ratio of 1:1:1:7. The mixture remains stable for 1 day

Dilute sarcolemmal vesicles from Protocol 5.3.2 to about 1 mg protein/ml with Soln. A (for protein concentration determination the Lowry method is recommended). Add 1 µl of Soln. B per 100 µl SL dilution, vortex and put on ice. Pipet the probes in triplicates into Eppendorf tubes according to Table 5.6. Label the ouabain-containing tubes with "+"; those without ouabain should be signed by "–".

Incubate the probes at 37 °C for 5 min, then start the enzymatic reaction by addition of 15 µl of SL dilution to each "+" and "–". Shake the probes at 37 °C and stop by addition of 0.5 ml Soln. D after exactly 15 min.

Determine the inorganic phosphate in a 0.5 ml aliquot of each sample (procedure see below). The specific activity of the oubain-sensitive Na,K-ATPase as well as the total ATPase activity is calculated from inorganic phosphate formed in the presence of ouabain

Table 5.6. Protocol for estimation of Na,K-ATPase activity

Solution	+ Ouabain	– Ouabain (µl)	Blank
C	375	375	375
ATP	50	50	50
Ou	50	–	–
ddH_2O	10	60	75
Pre-incubation at 37 °C for 5 min			
SL vesicle dilution	15	15	–

and absence of the drug, "P_i^+" and "P_i^-", blank B, and amount of used Protein "a" (mg) according to the equation

$$\mu\text{kat Na,K} - \text{ATPase/mg} = \frac{\mu\text{molP}_i^- - \mu\text{molP}_i^+}{0.5 \cdot a \cdot 900}$$

Determination of inorganic phosphate (see Protocol 1.3.2)
Mix 0.50 ml aliquots of the ATPase assay and phosphate standards (from 10 mM KH_2PO_4 (= 10 nMol/µl) in the range between 50 and 350 nmoles P_i per 0.5 ml) with 1.50 ml Soln. H. Close the reaction tubes and incubate at 37 °C in the dark for 1.5–2 h. Read at 750 nm and calculate from the standard curve.

5.3.2.2 Receptor Determination: DHP Binding Sites on Surface Membranes

Surface membranes of excitable cells contain a voltage-dependent L-type calcium channel, a membrane-spanning complex of several proteins. This complex is a valuable marker for this membrane type. It binds very specific calcium channel antagonists, such as, for instance, drugs of the dihydropyridine (DHP) type as (+)-PN 200–110 or nitrendipine.

> ***Important!*** *DHP's are light sensitive. Avoid daylight and work in a dark room illuminated by yellow light (e.g., sodium light source).*

Solutions/Reagents

A 50 mM Tris, pH 7.4, 2 mM $CaCl_2$, 0.1 mM PMSF (added just before use)
B PN: (+)-[Methyl-^{3}H]PN 200–110 (100 000 to 120 000 dpm = 1670 to 2000 Bq; specific radioactivity ca. 3 TBq/mmol) in 50 µl Soln. A (starting dilution)
C blank: 1 µM nitrendipine in Soln. A (final concentration)
D precipitating agent: 10% PEG 6000 (w/v) in 10 mM Tris, pH 7.4, 10 mM $MgCl_2$, kept in an ice bath
E 0.3% polyethylenimine (PEI) in pure water

Dilute a membrane preparation to about 0.4 mg protein per milliter or cell suspensions to about 10^6 cells/ml with Soln. A. Prepare four 1:3 dilution series of Soln. B using Soln. A as diluent (dilutions "0" to "4"). Pipet 50 µl of dilutions "0" to "4" as duplicates into scintillation vials to define the total radioactivity values "$Total_0$" to "$Total_4$."

First pipet Soln. A, B, and C according to Table 5.7 into disposable 4-ml tubes. Then add membrane or cell suspension to start the ligand binding and incubate at 37 °C for 1 h. Put the tube into an ice bath and stop by addition of 2 ml each of ice-cold Soln. D.

Filter the samples on Whatman GF/C glass fiber filters, wetted with Soln. E, in a holder manifold and wash twice with 2 ml of ice-cold Soln. D. Transfer the filters into scintillation vials, add scintillation cocktail, and count for radioactivity.

Table 5.7. Protocol for the determination of ligand (PN) binding

	Soln. A	Soln. B	Soln. C (µl)	Membrane and cell suspension
$Total_0$	–	50	–	–
$Blank_0$	50	50	50	100
$Bound_0$	100	50	–	100
$Total_1$	–	50	–	–
$Blank_1$	50	50	50	100
$Bound_1$	100	50	–	100
$Total_2$	–	50	–	–
$Blank_2$	50	50	50	100
$Bound_2$	100	50	–	100
$Total_3$	–	50	–	–
$Blank_3$	50	50	50	100
$Bound_3$	100	50	–	100
$Total_4$	–	50	–	–
$Blank_4$	50	50	50	100
$Bound_4$	100	50	–	100

5.3.2.3 Determination of the Dissociation and Association Kinetics of the DHP Receptors

A more detailed analysis of ligand binding is possible by determination of the association rate constant ("on-kinetics") of the dissociation rate constant ("off-kinetics"). The methodology of these estimations is illustrated for the cardiac DHP receptor.

Solutions/Reagents

- A 50 mM Tris, pH 7.4, 2 mM $CaCl_2$, 0.1 mM PMSF (added just before use)
- B (+)-[Methyl-^{3}H]PN 200–110 (60 000 to 80 000 dpm = 1000–1340 Bq; specific radioactivity ca. 3 TBq/mmol) per milliter of Soln. A
- C blank: 6.7 µM nitrendipine in Soln. B
- D precipitating agent: 10% PEG 6000 (w/v) in 10 mM Tris, pH 7.4, 10 mM $MgCl_2$, kept in an ice bath
- E 0.1 mM nitrendipine in Soln. A

Off-kinetics

For blank value mix 2×10^6 cells in 850 µl of Soln. A with 150 µl Soln. C and incubate at 37 °C for 60 min. Precipitate with Soln. D and filter on glass fiber filter as described above (t = 0; → "B_0").

Simultaneously suspend 2×10^7 cells in 10 ml of Soln. B and incubate at 37 °C for 60 min. Add Soln. E to a final concentration of 1 µM after this pre-incubation, vortex, and continue incubation. Take two aliquots of 50 µl each at t_0 (time of nitrendipine addition) and 1, 2, 5, 7, 10, 15, 20, 25, and 30 min after nitrendipine addition, precipitate with 2 ml of ice-cold Soln. D and suck off on glass fiber filter filters (→ "B_t"). Count for radioactivity and plot B_t/B_0 vs time t.

On-kinetics

Dilute 2×10^7 cells in 17 ml Soln. A and warm up to 37 °C. Prepare a solution of ^{3}H-PN 200–110 (about 1 600 000 dpm) in 3 ml. Mix both solutions and continue incubation at 37 °C. Take 1-ml probes in duplicate at time t = 0, 1, 2, 4, 6, 8, 10, 12, 15, and 20 min after mixing (→ "B_t"). Precipitate the probes with 2 ml of ice-cold Soln. D immediately after sampling and filter on Whatman GF/C glass fiber filters. Prepare a blank ("B_0") by incubation of a mixture of 850-µl cell suspension and 150 µl Soln. C at 37 °C for 20 min. Wash the filters with Soln. D. Dry the filters at air and count for radioactivity.

Plot B_t/B_0 vs t and calculate the maximal number of binding sites B_{max} and the dissociation constant K_D using either a computer program (e.g., Prism) or a SCATCHARD plot (see Chap. 9.2).

If B/F (B, bound ligand PN: $B = B_t - B_0$; F, free ligand PN: F = total PN – B) is plotted vs F, a straight line of the function

$$\frac{B}{F} = a + b \cdot B$$

is obtained. The x-axis intercept (y = B/F = 0) represents B_{max}; the slope b gives K_D.

The association rate constant k_1 is calculated from the slope of the function $B/B_0 = f(t)$, the dissociation rate constant k_{-1} is the slope of the equation

$$\ln(B/B_0) = -k_{-1} \cdot t$$

(for details see Chap. 9.2.1)

References

Glossmann H, Striessnig J (1990) Rev Physiol Biochem Pharmacol 114:1

5.3.3 RNA Separation by Non-Denaturating Sucrose Density Gradient Centrifugation

Solutions/Reagents

A 1 M NaCl, 0.1 M Tris · HCl, 5 mM EDTA, pH 7.5
B 15% sucrose (RNase free) (w/v) in Soln. A[3]
C 30% sucrose (RNase free) (w/v) in Soln. A

Prepare a linear gradient from Soln. B and Soln. C in a centrifuge tube (e.g., 13 ml Ultra-Clear for Beckman SW 41 rotor; 6 ml of Soln. B and Soln. C each) at the day before run. If a gradient mixer is used, e.g., according to Fig. 2.2, place the centrifuge tube either in a slanting position, start with Soln. C, and let flow along the tube

[3] Boil Soln. B and C in the presence 0.5% diethylpyrocarbonate (v/v) on the day before use.

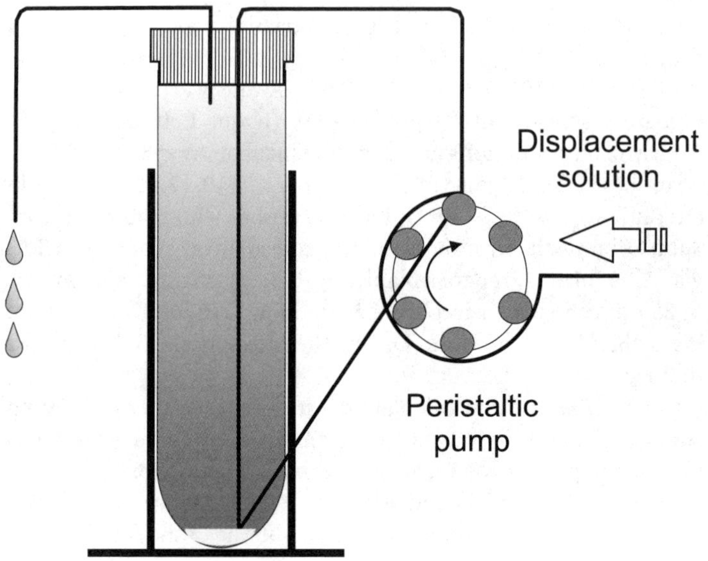

Fig. 5.4. Upward displacement gradient unloader

wall, or locate the tube vertically, fix the outlet tubing of the gradient mixer to the bottom of the centrifuge tube, and start the gradient with Soln. B. Store the filled tubes in a refrigerator overnight (avoid any vibrations).

Cover the sucrose gradient with 0.10 ml RNA-containing sample and precisely tare the tubes which will be opposite in the rotor.

Spin the samples with 200 000 × g_{max} at 4 °C for 15–17 h. After the run, displace the gradient using a more dense solution, e.g., 40% sucrose in Soln. A, colored by a droplet Amido Black 10 B solution. The principle of a displacement apparatus is shown in Fig. 5.4. The RNA content of the fractions is measured either by reading the UV absorption at 260 nm or, if labeled material was used, by counting the radioactivity. To monitor the sucrose gradient, estimate the refractive index of the obtained fractions (concentration, density and refractive index of sucrose solutions are given in Table 8.17).

5.3.4 Denaturating RNA Gradient Centrifugation

Solutions/Reagents

A mix 1 ml of 1 M HEPES with 99 ml of high-quality DMSO[4]
B 2.5% sucrose (w/v) in Soln. A
C 5.0% sucrose (w/v) in Soln. A
D 7.5% sucrose (w/v) in Soln. A
E 10.0% sucrose (w/v) in Soln. A
DMSO (ultra pure)
DMF (ultra pure)

[4] Distil DMSO in a nitrogen atmosphere; b.p.$_{18}$ 86 °C, b.p.$_{8}$ 63 °C, n_D^{20} 1.4787

Fill 1/5 of the effective volume of a centrifuge tube with Soln. B, cover it with the same volume of Soln. C, followed by Soln. D and Soln. E. Store the centrifuge tubes vibrationless at RT overnight. A linear gradient is formed by diffusion during this time.

Mix 60 µl of RNA solution with 275 µl DMSO, followed by 165 µl DMF. Cover each gradient with 200 µl of this RNA solution.

Spin the samples with 200 000 × g_{max} at 25 °C for 15–17 h.

RNA monitoring by UV reading is not possible, because DMSO absorbs below 275 nm, i.e., within the range of the nucleic acid absorption.

References

Rickwood D (ed.) (1992) Preparative centrifugation: a practical approach. Oxford University Press, London

5.3.5 Isopycnic Centrifugation

Isopycnic (buoyant density, equilibrium) separations should be performed in fixed angle or vertical rotors, because swing-out rotors give relatively large differences in centrifugal fields, i.e. the difference between g_{min} and g_{max} is much larger than in fixed angle rotors. If available, vertical rotors should be preferred, because they give shorter separation times than fixed angle rotors if run at high speed.

If self-forming cesium salt gradients are used, take care to be within the solubility of the salt at the given temperature to avoid crystallizations during centrifugation, since salt crystals may destroy the rotor during run (solubilities of cesium chloride and sulfate are given in Table 5.4).

The maximal speed of a rotor has to be reduced if solutions with densities $\varrho > 1.2$ g/ml are used. The proper revolutions per minute N_{red} are calculated by

$$N_{red} = N_{max} \cdot \sqrt{\frac{1.2}{\varrho}}$$

5.3.5.1 Purification of High Molecular Weight DNA in CsCl Gradients

This protocol is an example of separation using a self-generating density gradient.

Solutions/Reagents

A 10 mM Tris · HCl, 5 mM EDTA, pH 7.8
cesium chloride (ultrapure)

Dissolve 2.0 g cesium chloride (ultrapure) in 2.0 ml of Soln. A, containing 5–8 µg DNA. Stir with a glass rod at 20 °C until the salt is completely dissolved. Pour the solution into a centrifuge tube, e.g., for Beckman Coulter NVT 65.2 rotor, and centrifuge at 350 000 × g (18 900 rpm) at 20 °C for 6 h.

After the run, unload the tube either by puncture of the thin-walled tube with a syringe needle at the bottom, or by displacement by 80% cesium chloride (w/w). For DNA content, read UV absorption of the fractions and refractive index for density calculation (for equations see footnote to Table 5.4).

The content of guanine G and cytosine C of double-stranded DNA is calculated from density according to SCHILDKRAUT et al.:

$$\text{Mol} - \%\text{GC} = \frac{(\varrho - 1.66) \cdot 100}{0.098}$$

References

Osterman LA (1984) Methods of protein and nucleic acid research, vol. 1: Electrophoresis, isoelectric focusing, ultracentrifugation. Springer, Berlin, p. 284

Schildkraut CL, Marmur J, Doty P (1962) J Mol Biol 4:430

5.3.5.2 Cell Fractionation Using Percoll

Percoll is a polyvinylpyrrolidone-coated silica sol with very low osmolarity. For cell separations therefore the osmolarity is adjusted to isotonic conditions by mixing 9 vol. Percoll stock suspension with 1 vol. of 2.5 M sucrose, 1.5 M sodium chloride, or tenfold concentrated cell culture medium (e.g., 10 × RPMI).

Prepare the density gradient either stepwise or continuously; generation of the gradient by centrifugation is also possible. In comparison with cesium chloride gradients, the Percoll gradient forms fast during centrifugation.

Solutions/Reagents

- A 2.5 M sucrose or 1.5 M NaCl or tenfold concentrated cell culture medium
- B 1 vol. Soln. A and 9 vol. ddH_2O
- C Percoll working dilution: dilution of Percoll with a density nearby the density of the cells to be isolated; to prepare V ml of working dilution mix V/10 ml of Soln. A with $V_{0\,ml}$ Percoll stock[5] and $(V - V_0)$ ml ddH_2O

For preparation of a density gradient with a large linear range, manufacturers give the following conditions: Spin Percoll in 0.15 M NaCl (final concentration) with at least 10 000 × g_{av}, for Percoll in 0.25 M sucrose apply 25 000 × g_{av}. Use near vertical or vertical rotors for cell separation.

[5] $V_0 = V \cdot (\varrho - 0.1 \cdot \varrho_A - 0.9)/\varrho_0$; V_0, ml Percoll, V, final volume of working dilution, ϱ, required density of working dilution, ϱ_0, density of Percoll stock solution (≈ 1.13 g/ml), ϱ_A, density of Soln. A (1.5 M NaCl: 1.058 g/ml; 2.5 M sucrose: 1.316 g/ml)

The sample should contain about 8×10^7 cells in 1-ml culture medium per 12 ml total gradient volume and 0.5 ml sample (2–10 mg protein per milliliter) per 10 ml gradient volume. If self-generating gradients are used, separation is done with a $g \cdot t$-value of $1.35 \times 10^6\ g \cdot min$ (30 000 rpm, 45 min), in case of pre-formed gradients run with 800 g_{av} for 15 min. Allow the rotor to decelerate without brake to avoid swirling of the bands.

Collect the fraction containing the wanted cells. Dilute this fraction with a fivefold volume of Soln. B and harvest the cells by centrifugation at 200 × g for 20 min. Wash once with Soln. B.

Pelleting Percoll

Free subcellular particles or virus from Percoll by centrifuging the respective fraction of cell fractionation in a fixed-angle rotor at 100 000 × g for 90 min. Percoll forms a solid pellet which contains practically no material of interest.

References

Amersham Biosciences (2001) Percoll: Methodology and applications. Edition AC. Amersham Biosciences AB, Uppsala; www.amershambiosciences.com

Graham J (2001) Biological centrifugation. BIOS, Oxford, p 85

5.3.5.3 Preparation of Human Lymphocytes

Solutions/Reagents

- A 3.8% sodium citrate (w/v) in ddH_2O
- B RPMI medium, 10 mM HEPES[6]
- C RPMI medium, 5% FCS (v/v), 0.35 ml/100 ml 2-mercaptoethanol, 10 U/100 ml streptomycin, 10 U/100 ml penicillin, 29 mg/100 ml L-glutamine
- D IMDM medium, 10% human AB serum (v/v), 0.35 ml/100 ml 2-mercaptoethanol, 10 U/100 ml streptomycin, 10 U/100 ml penicillin, 29 mg/100 ml L-glutamine
- E 0.5% Trypan blue (C.I. 23850) in PBS

Important: *All solutions have to be filtrated sterile.*

Mix 10 ml freshly collected venous peripheral blood intensively with 2.5 ml Soln. A at RT. Then add 12.5 ml Soln. B, warmed to RT. Carefully cover the top of 25 ml of Lymphoprep (ϱ = 1.077 g/ml) with the diluted blood and spin with 800 × g (about 1600 rpm) at 20 °C for 30 min.

Discard the supernatant and harvest the cells. Dilute the cell suspension with the same volume of Soln. B and centrifuge at 20 °C with 500 × g (1000 rpm) for 10 min. Dilute the sediment with 30 ml of Soln. B and spin again.

---

[6] RPMI, modified McCoy 5 A medium for cell culture; HEPES, 4-(2-hydroxyethyl)piperazine-1-ethanesulfonic acid; IMDM, Iscov's modified Dulbecco's medium; FCS, fetal calf serum

Suspend the sediment in 10 ml of Soln. C and centrifuge at 4 °C with 1500 rpm for 5 min. Repeat this step until the lymphocytes are free from thrombocytes. Dilute the lymphocyte concentrate with about 20 ml Soln. D for further experiments.

For determination of yield, mix 10 µl of the cell suspension with 10 µl Soln. E and count using an NEUBAUER ruling hemocytometer.

6 Radioactive Labeling

Many of radioactive isotopes are very useful for the following biochemical processes (Table 6.1). The radioactive label is introduced into macromolecules, especially proteins, either during biosynthesis, e.g., during translation in the presence of ^{35}S-methionine, or enzymatically, e.g., by use of ^{32}P-labeled ATP during protein phosphorylation by protein kinases, or chemically by modification of amino acid side chains. Examples for reagents used in chemical radiolabeling of proteins are given in Table 6.2.

The main advantages of radioisotopes are the simple and very sensitive detection of the labels as well as the fact that labeled groups are chemically identical to their naturally occurring analogues; thus, macromolecules may be monitored within their natural environment and without significant loss of material. But even at low doses radioactive isotopes are dangerous, especially when they are incorporated into biologically active molecules. Protection and disposal are the main disadvantages; therefore, non-radioactive labeling, such as biotinylation or fluorescent labeling, should be preferred whenever it is possible (examples given in Protocol 3.6.7 and Protocol 4.1.10, respectively).

Important! *Take whatever actions are necessary to ensure that you comply with the national or state regulations governing the use of radioactive materials! Follow the code of good laboratory practice in addition to specific precautions relating to the particular radionuclides used. Laboratory overalls, safety glasses, and surgical gloves must be worn at all times. Avoid radioactive pollution of your bench to avoid endangering of your surroundings.*

Golden rules for handling radioactive materials

Working with radioactivity is not more dangerous than working with infectious or poisonous material, especially if "ten golden rules"[1] are observed: 1. Understand the nature of the hazard and get practical training. 2. Plan ahead to minimize time spent handling radioactivity. 3. Distance yourself appropriately from sources of radiation. 4. Use appropriate shielding for the type of radiation. 5. Contain radioactive materials within defined work areas. 6. Wear appropriate protective clothing and dosimeters. 7. Monitor the work area frequently for contamination. 8. Follow the local rules

[1] Amersham Biosciences (2002) "Safe and secure – a guide to working safely with radiolabeled compounds"

Table 6.1. Biochemically important radioisotopes. Half-life, β-decay energy, and specific radioactivity

Radio-nuclide	Half-life				β Energy (fJ)	Specific activity (TBq/milliatom)
^{3}H	12.3	Years	108 887	Hours	3.0	1.07
^{14}C	5370	Years			25.1	0.0023
^{22}Na	2.6	Years	22 794	Hours	87.4	5.07
^{32}P	14.3	Days	43.3	Hours	273.6	336.2
^{33}P	25.4	Days	609.6	Hours	39.8	187.8
^{35}S	87.4	Days	2097.6	Hours	26.9	55.0
^{36}Cl	3.01×10^{5}	Years			113.4	43.9
^{45}Ca	163	Days	3912	Hours	41.1	29.5
^{51}Cr	27.7	Days	664.9	Hours	[a]	173.8
^{57}Co	271.8	Days	6523.2	Hours	[a]	17.7
^{58}Co	70.8	Days	1699	Hours	[a]	68.2
^{59}Fe	44.5	Days	1068.7	Hours	74.6	108.0
^{86}Rb	18.6	Days	446.4	Hours	283.8	258.5
^{125}I	59.9	Days	1437.6	Hours	[a]	80.3
^{131}I	8.05	Days	193.2	Hours	97.0	601.0

[a] Electron capture

and safe ways of working. 9. Minimize accumulation of waste and dispose of it by appropriate routes. 10. After completion of work, monitor yourself, wash, and monitor again.

The examples for radioactive labeling by phosphorus-32 (^{32}P) and iodine-125 (^{125}I) in this chapter were chosen for two reasons: on one hand, they are relatively easy to do, and on the other hand, the measurement of radioactivity is simple. ^{32}P is counted in water in a liquid scintillation counter by measuring the CERENKOV radiation and ^{125}I is measured in a gamma counter. Both isotopes may be detected also by autoradiography. A further advantage of both isotopes is their short half-life, which eases the disposal of nuclear waste.

6.1 Radioactive Decay

Law of radioactive decay:

$$N = N_0 \cdot e^{\frac{-t \cdot \ln 2}{t_{1/2}}}$$

N, radioactivity at time t; N_0, radioactivity at $t = 0$; $t_{1/2}$, half-life of the isotope.

Mass of a radioactive compound per MBq:

$$M = M_r \cdot (\text{kBq}) \cdot t_{1/2} \cdot C$$

M: mass of the radioactive compound in micrograms; M_r: relative molar mass and atom mass, respectively; (kBq): actual radioactivity of the compound in kBq; C: constant (cf. Table 6.3).

Specific radioactivity S at F percent isotopic purity ($t_{1/2}$ is given in days) is calculated by:

$$S = F \cdot 4.81 \cdot 10^{11}/t_{1/2} \quad \text{in MBq/mmol}$$

$$S' = S/M_r \quad \text{in MBq/mg}$$

Table 6.2. Reagents for radioactive labeling of proteins

Acceptor group within the protein	Labeling reagent (carrying a radioactive isotope)
$-NH_2$	1-Fluoro-2,4-dinitrobenzene
	Acetanhydride
	Bolton–Hunter reagent
	Dansylchloride
	Formaldehyde
	Maleic acid anhydride
	N-Succinimidyl propionate
	Phenylisocyanate
	Sodium borohydride
	Succinic anhydride
-SH	Acetanhydride
	Bromoacetic acid and derivatives
	Chloroacetic acid and derivatives
	Dansylchloride
	Iodoacetamide
	N-Ethylmaleinimide
	p-Chlormercuribenzoic acid
	Acetanhydride
Phenyl, hydroxyphenyl	Dansylchloride
	Iodine
Histidiyl	Dansylchloride
	Iodine
$-CH_2OH$	Acetanhydride
	Diisopropyl fluorophosphate

[a] The reagents may be labeled by different radioactive isotopes, e.g., ^{3}H, ^{14}C, or ^{125}I

6.2 Decay Tables for 32-Phosphorus, 35-Sulfur, and 125-Iodine

The number N in Tables 6.4–6.6 represents that part of radioactivity present at day n after starting the experiment.

Table 6.3. Factors for calculation of specific radioactivity

$t_{1/2}$	C
Seconds	3.285×10^{-12}
Minutes	1.971×10^{-10}
Hours	1.180×10^{-8}
Days	2.834×10^{-7}
Years	1.036×10^{-4}

Table 6.4. ^{32}Phosphorus decay table

Days	Number	Days	Number	Days	Number
−5	1.274	11	0.587	27	0.270
−4	1.214	12	0.559	28	0.257
−3	1.157	13	0.533	29	0.245
−2	1.102	14	0.507	30	0.234
−1	1.050	15	0.483	31	0.223
0	1.000	16	0.460	32	0.212
1	0.953	17	0.439	33	0.202
2	0.908	18	0.418	34	0.192
3	0.865	19	0.398	35	0.183
4	0.824	20	0.379	36	0.175
5	0.785	21	0.361	37	0.166
6	0.748	22	0.344	38	0.159
7	0.712	23	0.328	39	0.151
8	0.679	24	0.312	40	0.144
9	0.646	25	0.298	41	0.137
10	0.616	26	0.284	42	0.131

Table 6.5. ^{35}Sulfur decay table

Days	Number	Days	Number	Days	Number
−5	1.040	21	0.847	49	0.678
0	1.000	28	0.801	56	0.641
3	0.976	32	0.776	60	0.621
7	0.946	37	0.746	65	0.597
9	0.931	39	0.734	67	0.588
11	0.916	42	0.717	70	0.574
14	0.895	44	0.705	72	0.565
16	0.881	46	0.694	74	0.556
18	0.874				

Table 6.6. ^{125}Iodine decay table

Days	Number	Days	Number	Days	Number
−10	1.122	24	0.758	58	0.512
−8	1.097	26	0.741	60	0.500
−6	1.072	28	0.724	62	0.489
−4	1.047	30	0.707	64	0.477
−2	1.023	32	0.691	66	0.467
0	1.000	34	0.675	68	0.456
2	0.977	36	0.660	70	0.445
4	0.955	38	0.645	72	0.435
6	0.933	40	0.630	74	0.426
8	0.912	42	0.616	76	0.416
10	0.891	44	0.602	78	0.406
12	0.871	46	0.588	80	0.397
14	0.851	48	0.574	82	0.388
16	0.831	50	0.561	84	0.379
18	0.812	52	0.548	86	0.370
20	0.794	54	0.536	88	0.362
22	0.776	56	0.524	90	0.354

Examples

a) If started with 15 kBq ^{32}P, $15 \cdot 0.787 = 11.805$ kBq are present 5 days later.
b) When 8000 dpm^{32}P are measured at day 6, the sample originally contained 8000/0.748 = 10 695 dpm and 178.25 Bq.

6.3 Enzymatic [^{32}P]-Phosphate Incorporation into Proteins

Chemical stability of amino acid phosphates

A lot of biologic membrane systems and cellular organelles contain kinases, which transfer phosphate groups to proteins, especially to serine, threonine, and tyrosine residues. Self-phosphorylation of enzymes leading to acylphosphates or phosphoamides can be observed, too. With respect to their chemical stability, these phosphoproteins are classified into acid-stable (alkali labile), hydroxylamine-sensitive, and acid labile.

References

Weller M (1979) Protein phosphorylation: the nature, function, and metabolism of proteins with covalently bound phosphorus. Pion, London

The catalytic subunit of the cycloAMP-dependent protein kinase (cAMP PrK) transfers the terminal (γ) phosphate residue of ATP to serine residues of numerous proteins. The assay described in this protocol allows cAMP-dependent phosporylation as well as other enzymatic protein phosphorylations.

Solutions/Reagents

A 480 mM KCl, 160 mM histidine, 40 mM $MgCl_2$, pH 6.8
B 2.5 mM Tris-ATP[2,3]
C [γ-^{32}P]ATP in Soln. B 10–20 kBq/µl
D 0.5 M NaF in ddH_2O
E 2 mM EGTA
F 50 µg/ml catalytic subunit of cAMP PrK
G 50 µg/ml cAMP PrK inhibitor[4]
H 50 µg/ml calmodulin
I 2 mM $CaCl_2$
J 50 mM KCl, 20 mM Tris·HCl, 0.2 mM DTE or DTT, pH 6.8
K 15% trichloroacetic acid (w/v), 50 mM sodium phsophate in water

Pipet the assay according to Table 6.7, cool the tubes in an ice bath, then add the protein solution or membrane suspension which

Table 6.7. Protocol for phosphorylation with protein kinases (PrK)

Type of protein kinase	Solution						
	A	D	E	F	G	H	I
	(µl per 200 µl assay)						
Exogenous cAMP PrK	50	10	10	10	–	–	–
Control	50	10	10	–	10	–	–
Endogenous calmodulin-sensitive PrK	50	10	–	–	10[a]	10	10
Control	50	10	10	–	10[a]	–	–
Endogenous PrK in the presence of endogenous calmodulin	50	10	–	–	10	–	10
Control	50	10	10	–	10	–	–
Other endogenous PrK[b]	50	10	10	–	–	–	–
Control	50	10	10	–	10	–	–

[a] Omit the inhibitor of cAMP PrK if this enzyme is not present or if its activity is part of the control

[b] Potential activators of endogenous protein kinases, e.g. cycloAMP, cycloGMP, inositol phosphates, or lipids may be added in physiological concentrations

[2] Pour a solution of sodium ATP onto a small column of a cation exchange resin, equilibrated with Tris. Elute Tris-ATP with ddH_2O and monitor by UV absorption at 260 nm.

[3] If the specific radioactivity of the [γ-^{32}P]ATP is not very high or if endogenous phosphatases are absent, the final concentration of 250 µM ATP may be reduced.

[4] Preparation of cAMP PrK inhibitor according to Demaille JJ, Peters KA, Fischer EH (1977) Biochem 16:3080

should contain 10–200 µg of protein, and fill up with ddH_2O to 200 µl each. Pre-incubate at 30 °C for 3 min, and then start phosphorylation by addition of 20 µl of Soln. C.

Stop the phosphorylation: either by 50-fold dilution with ice-cold Soln. J and centrifugation at 0 °C with 50 000 to 100 000 × g for 30 min (in the case of membrane phosphorylation), or by addition of 1 ml of ice-cold Soln. K after 10 s to 5 min.

For determination of total ^{32}P-phosphorus incorporation, filter the precipitate over a glass fiber filter and count for radioactivity. If electrophoresis is intended, collect the precipitate by centrifugation with 5000 × g and dissolve the pellet in the appropriate sample buffer. Choose the electrophoresis system with respect to the different pH stability of phosphorylation products.

6.4 Iodination with [^{125}I]-Iodine Reagents

***Caution!** 125-Iodine may be liberated during iodination even when solutions are used. Work in a special hood! Check your thyroid gland after 6 and 24 h if you have worked with iodination reagents.*

Iodine becomes incorporated into proteins either oxidatively, or enzymatically, or electrochemically. Oxidative incorporation uses organochemical oxidants, such as, for example, chloramine T or Iodo-gen (1,3,4,6-tetrachloro-3α,6α-diphenyl glycouril). Enzymatic incorporation is done by means of lactoperoxidase. By these methods iodine is introduced into phenyl (tyrosyl) residues of the protein.

Conjugation of a protein with the radiolabeled Bolton–Hunter reagent in the first place modifies lysine side chains.

6.4.1 Chloramine-T Protocol

Solutions/Reagents

A Na^{125}I solution, 3.7 GBq/ml
B 0.25 M sodium phosphate buffer, pH 7.5
C 50 mM sodium phosphate buffer, pH 7.5
D 5 mg/ml chloramine-T (sodium N-chloro-p-toluenesulfonamide) in Soln. C
E 0.3% 2-mercaptoethanol (v/v) in Soln. C[5]
F 0.1% serumalbumin (w/v) or 0.1% gelatin (w/v) or 2% heat-inactivated serum (v/v) in Soln. C
G 0.2% NaI (w/v) in Soln. F

Prepare Soln. D, E, and G immediately before starting the experiment.

[5] To avoid reduction of disulfide bonds, use 0.2 mg/ml tyrosin instead of 2-mercaptoethanol.

Pipet 5–10 µl of Soln. A (4–7 MBq) into a 4-ml polystyrene or siliconized glass test tube. Then add 25 µl Soln. B, followed by 10 µl of protein solution (2–5 µg, dissolved in Soln. C), 10 µl Soln. D, and 100 µl Soln. E. Vortex after each addition. Fill up to 1 ml with Soln. G and count for total radioactivity.

Equilibrate a 1 × 10 cm Sephadex G-25 column with 50 ml Soln. F. Apply the iodinated sample, elute with Soln. F, and collect 1-ml fractions. The iodinated protein appears in the void volume, and free iodine-125 elutes with the total volume. Estimate its radioactivity and calculate the specific radioactivity (Bequerel per milligram of protein).

If the iodinated protein will be used in bioassays, check for biologic activity (antigenicity, enzymic activity, ligand binding, etc.) in comparison with the not-labeled starting material.

6.4.2 Iodination with Bolton–Hunter Reagent

Solutions/Reagents

- A Bolton–Hunter reagent (N-succinimidyl 3(4-hydroxy-5-[^{125}I]-iodophenyl)propionate) (70–150 TBq/mMol) in benzene
- B 0.1 M borate buffer, pH 8.5
- C 0.2 M glycine in Soln. B
- D 0.1% serumalbumin (w/v) or 0.1% gelatin (w/v) in 50 mM phosphate buffer, pH 7.5

Pipet about 5–10 MBq (about 250 000–500 000 dpm) of Soln. A into a 4-ml test tube. Carefully vaporize the solvent by a gentle stream of dry nitrogen. Add 10 µl of protein solution, containing 2–5 µg of protein in Soln. B, and agitate the tube in an ice bath for 30 min. Add 0.5 ml Soln. C and continue shaking at 0 °C. Fill up to 1 ml with Soln. D after an additional 5 min. Separate the labeled protein from the unreacted reagent as described in Protocol 6.4.1.

References

Bolton AE (1985) Radioiodination techniques (Amersham Review 18), 2nd ed., Amersham International plc, Amersham

Bolton AE, Hunter WM (1973) Biochem J 133:529

Hermanson GT (1996) Bioconjugate techniques. Academic Press, New York, p 414

6.5 Scintillation Cocktails for Liquid Scintillation Counting

The scintillation cocktail chosen depends on the radioisotope and the type and amount of solvent containing the labeled compound. High-energy β emitting isotopes, such as 32-phophorus, are detectable by several methods such as liquid scintillation (Cerenkov counting) in water, application of solid scintillators, autoradiography on X-ray films without enhancers, or special equipment (PhosphoImager). Much more difficult is the counting of weak emitters, such as, for example, tritium, since a lot of substances quench radiation.

A lot of ready-to-use scintillation cocktails are available. Besides counting efficiency, chemical, biologic, and environmental hazards should be reflected when choosing a cocktail. Since vials and their contents are radioactive waste, number and volume of vials should be as small as possible.

To illustrate the composition of liquid scintillation cocktails (LSC), some formulations follow.

A: Bray solution: 6% naphthalene (w/v), 0.4% PPO (2,5-diphenyloxazole) (w/v), 0.02% POPOP (1,4-bis-2-(5-phenyloxazolyl)-benzene) (w/v), 10% methanol (v/v), 2% ethyleneglycol (v/v) in 1,4-dioxane

B: 15% naphthalene (w/v), 0, 8% PPO (w/v), 0.06% POPOP (w/v), 10% ethylglycol (ethyleneglycol monoethylether) (v/v), 2% ethyleneglycol (v/v) in 1,4-dioxane

C: 6.5 g PPO, 0.15 g POPOP, 100 g naphthalene, 300 ml methanol, 500 ml 1,4-dioxane, 500 ml toluene

D: "Tritosol" (handles up to 23% water (v/v) without decrease in counting efficiency for tritium)[6]: 0.3% PPO (w/v), 0.02% POPOP (w/v), 3.5% ethyleneglycol (v/v), 14% ethanol (v/v), 25% Triton X-100 (v/v) in xylene

---

[6] Fricke U (1975) Anal Biochem 63:555

7 Buffers

7.1 Theoretical Considerations

Macromolecules, e.g., proteins, need a distinct structure within the aqueous surrounding to realize their biologic functions. This structure is stabilized inter-alia by ionic interactions between positively and negatively charged amino acid side chains and between these chains and other molecules. Optimal functionality needs a well-balanced ratio of charged residues. Each disorder of this ratio results in alterations up to complete denaturation.

Incomplete dissociation of amino acid side chains in proteins

Since the side chains of a distinct amino acid do not show complete dissociation at common concentrations, these acids (when forming an anion during dissociation) and bases (when forming a cation) are called weak acids and bases, respectively, as opposed to the completely dissociated strong acids and bases, e.g., hydrochloric acid, trifluoroacetic acid, sodium hydroxide, or triethylammonium hydroxide.

Weak acids and bases are not restricted to amino acids; many other compounds also behave similar in aqueous solutions. The dissociation into anion A^- and proton H^+ (in aqueous solution the dissociated protein is captured by a water molecule to form a hydronium ion H_3O^+) in the case of acids, and the addition of a proton to a base B to form the cation BH^+, is an equilibrium

$$AH + H_2O \leftrightarrows A^- + H_3O^+$$
$$B + H_3O^+ \leftrightarrows BH^+ + H_2O$$

This equilibrium can be described for dissociation by the thermodynamic equation:

$$K_D = \frac{k_{-1}}{k_1} = \frac{[A^-] \cdot [H^+]}{[AH]}$$

with the dissociation equilibrium constant K_D and the concentrations $[A^-]$, $[H^+]$, and $[AH]$ of the reaction partners. As a thermodynamic parameter, the dissociation constant K_D depends on temperature and therefore the dissociation is also temperature dependent.

In a more detailed view are factors of the equilibrium equation, not the concentrations, but the activities a_x, i.e., only a part of the total amount of a reactant:

$$a_{A^-} = f_{A^-} \cdot [A^-] \text{ resp. } a_{HA} = f_{HA} \cdot [HA] \text{ resp. } a_{H^+} = f_{H^+} \cdot [H^+]$$

The factor f depends on the analytical concentration of a reactant and becomes 1 at infinite dilution. Factor f is significantly lower than 1, too, if a reactant is dissolved in the presence of a neutral salt, such as NaCl or KCl.

Transformation of the equilibrium equation gives an expression of the H^+-concentration in solution (HENDERSON–HASSELBALCH equation):

$$pH = pK + \frac{a_{A^-}}{a_{HA}}$$

"p" symbolizes the negative common logarithm of the respective quantity. The value of K_D of pure water at 20 °C is 10^{-14}; therefore, a neutral solution, i.e., $[A^-] = [H^+]$ has pH = 7.

Since the factors f are known for only a couple of substances, only more or less rough empirical calculations are possible[1]. To illustrate the concentration dependency of the activity factor f, Table 7.1 lists it for some substances.

The increase of f during dilution explains the observation that a buffer solution alters its pH when diluted.

The pH of a buffered solution also changes when temperature is increased or decreased. Table 7.2 gives K_a, molar mass M_r as well as dpK_a/dT data for some substances often used in buffer solutions. "$-\Delta pH/grd$" means that pH increases when temperature decreases. For example, Tris buffer adjusted to pH 7.8 at 22 °C has pH 8.4 at 4 °C, or during a PCR cycle the pH can differ up to 1.5 units.

The change of pH at different temperature is illustrated for some buffer substances in Fig. 7.1.

Table 7.1. Dependence of activity coefficients on concentration

Ion	Activity coefficient f		
Concentration:	0.001 M	0.01 M	0.1 M
H^+	0.98	0.93	0.86
OH^-	0.98	0.93	0.81
$Acetate^-$	0.98	0.93	0.82
$H_2PO_4^-$	0.98	0.93	0.74
HPO_4^{2-}	0.90	0.74	0.45
PO_4^{3-}	0.80	0.51	0.16
$Citrate^-$	0.98	0.93	0.81
$Citrate^{2-}$	0.90	0.74	0.45
$Citrate^{3-}$	0.80	0.51	0.18

[1] Buffer formulations and calculations at www.liv.ac.uk/buffers/buffercalc.html; software at http://www.liv.ac.uk/~jse/software.html

Table 7.2. Dissociation constant pK (25 °C), dpK/dT, and molecular mass of buffers

Substance	pK	dpK/dT	M_r
Oxalate (oxalic acid) (pK_1)	1.27		90.0
Maleate (maleic acid) (pK_1)	2.00	0	116.1
Phosphate (phosphoric acid) (pK_1)	2.15	0.0044	98.0
Glycine (pK_1)	2.35	−0.002	75.1
Phthalate (phthalic acid) (pK_1)	2.95		166.1
Glycylglycine (pK_1)	3.12		132.1
Malate (malic acid) (pK_1)	3.40		134.1
Formiate (formic acid)	3.75	0	46.0
Barbituric acid (pK_1)	4.04		128.1
Succinate (succinic acid) (pK_1)	4.21	−0.0018	118.1
Oxalate (oxalic acid) (pK_2)	4.27		90.0
Acetate (acetic acid)	4.75	0.0002	60.0
Citrate (citric acid) (pK_2)	4.76	−0.0016	192.1
Malate (malic acid) (pK_2)	5.13		134.1
Pyridine	5.23	−0.014	79.1
Phthalate (phthalic acid) (pK_2)	5.41		166.1
Piperazine (pK_2)	5.55	−0.015	86.1
Succinate (succinic acid) (pK_2)	5.64	0	118.1
Malonate (malonic acid) (pK_2)	5.70		104.1
Histidine (pK_2)	5.97	−0.017	155.2
Maleate (maleic acid) (pK_2)	6.07		116.1
MES (2-morpholinoethanesulfonic acid)	6.10	−0.011	195.2
Cacodylate (cacodylic acid. dimethylarsinic acid)	6.27		138.0
Carbonate (carbonic acid) (pK_1)	6.35	−0.0055	62.0
Citrate (citric acid) (pK_3)	6.40	0	192.1
BIS-Tris (bis(2-hydroxyethyl) imino-tris(hydroxymethyl)-methane)	6.46	−0.02	209.5
ADA (N-carbamoylmethyl-iminodiacetic acid)	6.59	−0.011	190.2
PIPES (1.4-piperazine-bis-ethane sulfonic acid)	6.76	−0.0085	302.4
ACES (N-(2-acetamido)-2-aminoethanesulfonic acid)	6.78	−0.02	182.2
BIS-Tris Propane (1.3-bis[tris(hydroxy-methyl)- methylamino]-propane) (pK_1)	6.80	−0.028	282.3
Imidazole	6.95	−0.02	68.1
MOPSO (3-(N-morpholino)-2-hydroxypopanesulfonic acid)	6.95	−0.015	225.3
BES (2-[bis(2-hydroxyethyl)-amino]-ethansulfonsäure)	7.09	−0.016	213.3
Phosphate (phosphoric acid) (pK_2)	7.21	−0.0028	98.0

Table 7.2. (continued)

Substance	pK	dpK/dT	M_r
MOPS (4-morpholino-propane sulfonic acid)	7.31	−0.011	209.3
TES (2-[tris(hydroxymethyl)-methyl-amino]-1-ethanesulfonic acid)	7.40	−0.02	229.5
HEPES (4-(2-hydroxyethyl)-1-piperazine-ethanesulfonic acid)	7.48	−0.014	238.3
DIPSO (3-[bis(2-hydroxyethyl)amino]-2-hydroxypropane-1-sulfonic acid)	7.60	−0.02	243.3
Triethanolamine (TEA)	7.76	−0.02	149.2
POPSO (piperazine-N.N′-bis(2-hydroxypropanesulfonic acid))	7.85	−0.013	362.4
Barbital (diethylbarbituric acid, Veronal)	7.98		184.2
HEPPS (EPPS. 4-(2-hydroxyethyl)-1-piperazin-propanesulfonic acid)	8.00		252.3
Tricine (N-[tris(hydroxymethyl)-methyl]-glycine)	8.05	−0.021	179.2
Glycinamide hydrochloride	8.06	−0.029	110.5
Tris (tris(hydroxymethyl)-amino methan; 2-amino-2-hydroxy methyl-propane-1.3-diol; Trizma)	8.06	−0.028	121.1
PIPPS (1.4-bis(3-sulfopropyl)-piperazine)	8.10		330.4
Glycylglycine (pK_2)	8.25	−0.025	132.1
Bicine (N.N-bis(2-hydroxyethyl)-glycine)	8.26	−0.018	163.2
TAPS (3-tris(hydroxymethyl)-methylamino)-1-propanesulfonic acid	8.40	0.018	243.3
AMPSO (3-([1.1-dimethyl-2-hydoxyethyl]amino)-2-hydroxypropane sulfonic acid)	8.40		227.3
Diethanolamine	8.88		105.1
Histidine (pK_3)	8.97		155.2
BIS-Tris propane (1.3-bis[tris (hydroxymethyl)-methylamino] propane) (pK_2)	9.00	−0.028	282.3
AES (2-aminoethylsulfonic acid; taurine)	9.06	−0.022	125.2
Borate (boric acid. H_3BO_3)	9.23	−0.008	61.8
Ammonia	9.25	−0.031	17.0
Ethanolamine	9.50	−0.029	61.1
CHES (2-cyclohexylaminoethanesulfonic acid)	9.55	−0.018	207.3
CAPSO (3-cyclohexylamino-2-hydroxypropanesulfonic acid)	9.71	−0.018	237.3
Glycine (pK_2)	9.78	−0.025	75.1
EDPS (N.N′-bis(3-sulfopropyl)-ethylenediamine)	9.80		
Carbonate (carbonic acid) (pK_2)	10.33	−0.009	62.0

Table 7.2. (continued)

Substance	pK	dpK/dT	M_r
CAPS (3-(cyclohexylamino)-1-propanesulfonic acid)	10.40	0.032	221.3
Ethylamine	10.62		45.1
Triethylamine	10.72		101.2
Piperidine	11.12	−0.031	85.2
Phosphate (phosphoric acid) (pK_3)	12.33	−0.026	98.0

Important: *Consider concentration and temperature dependency of buffering substances when you prepare and use a buffered solution.*

The capability to stabilize a given pH by a buffer depends on its capacity. Figure 7.3 shows as an example for Tris buffer, that the buffer capacity β (amount of acid or base which causes only a little change in pH) is highest in the range of ± 0.5 units around the pK and that e.g. at one pH unit distant from pK addition of a small amount of acid or base causes a huge change in pH.

Figure 7.2 illustrates that the ability of a buffering substance (this example: 0.1 M Tris) is highest at pH = pK and decreases with distance to pK.

The dissociation constant of a weak acid or base is altered by ionic strength, too. Ionic strength I is a measure for total content of ions within a solution and depends on concentration as well as on charge of ions:

$$I = \frac{1}{2} \cdot \sum_{i=1}^{n} (c_i \cdot z_i^2)$$

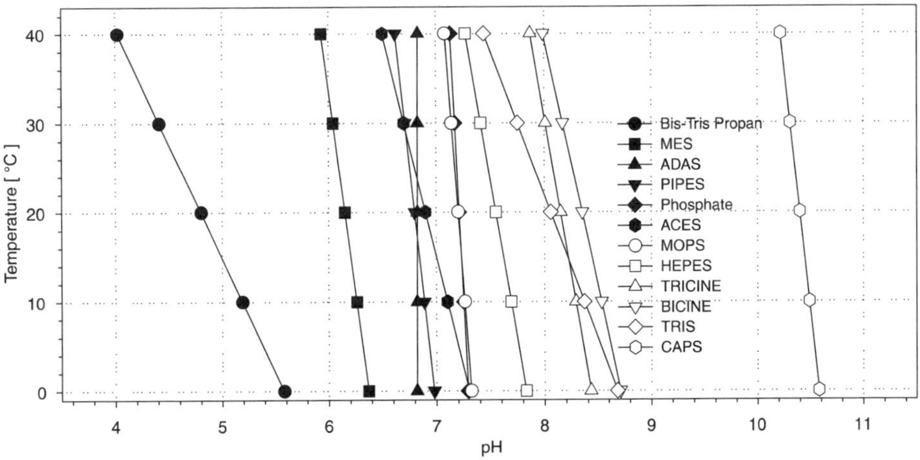

Fig. 7.1. Change of pH of some 0.1-M buffers depending on temperature

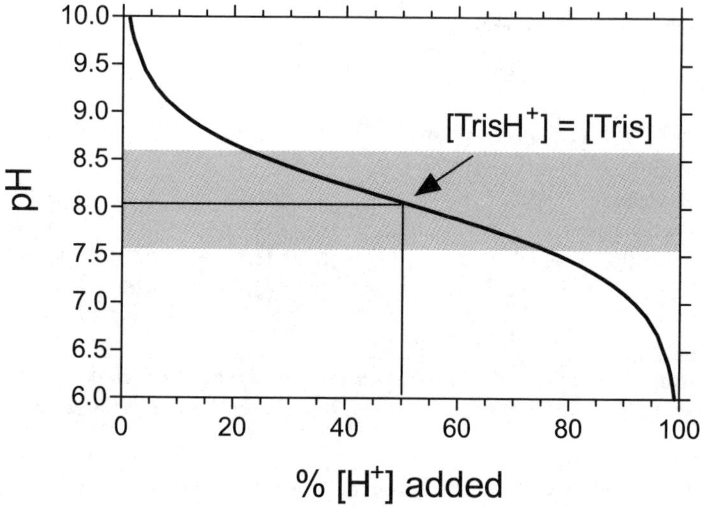

Fig. 7.2. Buffer capacity (titration curve; this example: addition of acid) for 0.05-M Tris buffer ($pK_{Tris} = 8.06$)

I: ionic strength; c_i: concentration of ion i; z_i: Charge of ion i; n: number of ion species within the solution.

Taking ionic strength into consideration, the modified dissociation constant pK′ may be calculated by the Debye-Hückel relationship:

$$pK' = pK + (2z_i - 1) \cdot \left[\frac{A \cdot \sqrt{I}}{1 + \sqrt{I}} - c_i \cdot I \right]$$

The term A is a temperature-dependent constant. Values of A between 0 and 60 °C are given in Table 7.3.

Since electrolytes, especially weak electrolytes, are not completely dissociated at higher concentrations, the degree of dissociation α_i of an ion has to be introduced into calculations (theoretically, a salt is totally dissociated at infinite dilution, i.e., at infinite

Table 7.3. Values of constant A at different temperatures

°C	A
0	0.4918
10	0.4989
20	0.5070
25	0.5114
30	0.5161
37	0.5321
40	0.5262
50	0.5373
60	0.5494

dilution $\alpha_i = 1$). The definition of ionic strength I with respect to incomplete dissociation is

$$I = \frac{1}{2} \cdot \sum_{i=1}^{n} \left(c_i \cdot \alpha_i \cdot z_i^2\right)$$

The degree of dissociation is calculated by

$$\alpha_i = \frac{10^{-pK}}{10^{-pK} + 10^{-pH}}$$

pK: dissociation constant; pH: value of pH for the respective buffer.

Since proteins contain a lot of acidic and basic side chains acting as weak acids and bases, respectively, proteins are buffering substances, too. If you mix buffer solutions with protein solutions, pH may be altered because the concentration of protein's buffering residues may exceed the capacity of the (chemical) buffer. For instance, bovine serum albumin contains 59 basic (Lys) and 99 acidic (59 Asp plus 40 Glu) residues per mole; a solution of 10 mg/ml (1%) BSA contains 9 mM basic and 14.5 mM acidic residues, and phosphate-buffered saline (PBS) contains only 10 mM phosphate. As a consequence of this example (a) the concentration of the chemical buffer should be high enough to act as a buffer, (b) choose a chemical buffer the pK of which is nearby the pH to be stabilized, and (c) adjust the pH after all components are mixed.

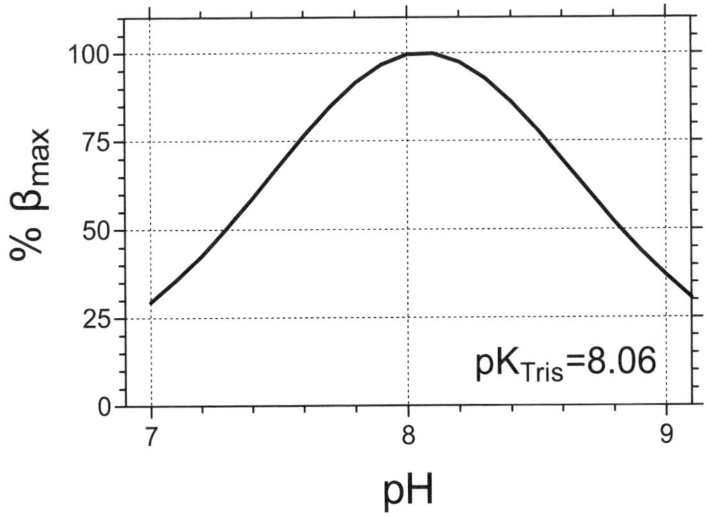

Fig. 7.3. Buffer capacity β depending on distance of pH to pK

References

Stoll VS, Blanchard JS (1990) Meth Enzymol 182:24
Beynon RJ, Easterby JS (1996) Buffer solutions. IRL Press, Oxford

7.2 Plot for Buffer Calculations

Figure 7.4 is the graphical expression of the HENDERSON–HASSELBALCH equation. The graph may be used for estimation of buffer composition, if the pK of the buffering substance is known. For example, the base A^- of an acetate buffer is the anion CH_3COO^-, the acid HA is CH_3COOH, and in a Tris buffer, $TrisH^+$ is the acid HA and unprotonated Tris is the base A^- (according to the BRÖNSTED definition, the acid is that substance which has bound the hydrogen cation):

$$pH = pK + \frac{[A^-]}{[HA]}$$

Abscissa — Difference between wanted pH and pK

Ordinate — Quotient of concentrations of base and acid (e.g., $HPO_4^-/H_2PO_4^-$)

Examples

i. wanted: Tris buffer, pH 8.2, 0.1 M, 4 °C
   given: $pK_4 = 8.51$
   solution: $\Delta pH = (8.2 - 8.51) = -0.31$

The abscissa value $\Delta pH = -0.31$ in Fig. 7.4 gives $[A^-] : [HA] = 0.48$, i.e., the ratio of Tris base to Tris · HCl is 0.48.

$$0.48 = (\text{moles Tris} - \text{moles HCl}) : \text{moles HCl} = (0.1 - x) : x$$

From this follows: $x = 0.0676$

To prepare the wanted buffer, dissolve 12.11 g Tris (0.1 mol) in about 900 ml water, add 67.6 ml 1 N HCl (0.0676 moles), and bring up to 1000 ml.

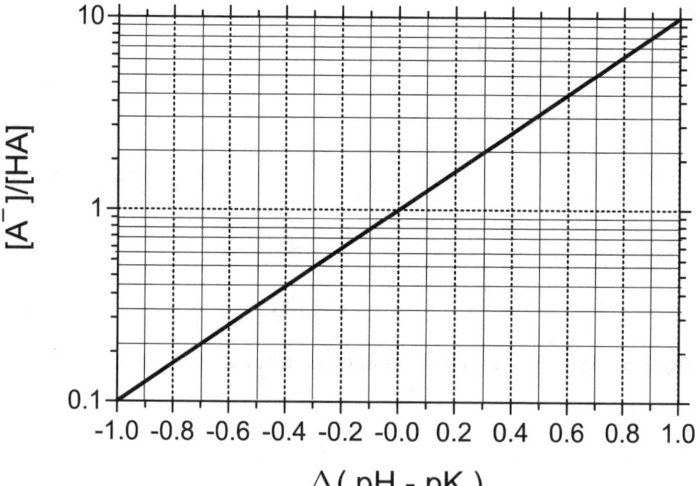

Fig. 7.4. Graph for buffer estimation

ii. wanted: acetate buffer, pH 4.96, 0.1 M, 20 °C
given: $pK_{20°} = 4.76$
solution: $\Delta pH = (4.96 - 4.76) = 0.2$

For abscissa 0.2, the ordinate $[A^-] : [HA]$ is 1.6.

If acetic acid and sodium hydroxide solution are used for preparation of the buffer, the amounts are calculated as follows:

$$1.6 = (\text{moles NaOH}) : (\text{moles } CH_3COOH - \text{moles NaOH})$$
$$= x : (0.1 - x)$$

From this follows: $x = 0.061$

Mix 0.1 mol/l acetic acid with 0.061 mol/l sodium hydroxide to get the expected buffer.

If using acetic acid and sodium acetate, the following results are obtained:

$$1.6 = (\text{moles Na acetate}) : (\text{moles acetic acid}) \quad \text{and}$$
$$0.1 = (\text{moles Na acetate}) + (\text{moles acetic acid})\ .$$

Per liter buffer, 0.0615 mol sodium acetate and 0.0385 mol acetic acid are needed.

7.3 pH Indicators

Solutions of indicators are out for pH measuring, but they are useful for monitoring some processes, such as tissue cultures (indicator: phenol red) or electrophoresis (electrophoresis front and pH of sample; indicator: bromophenol blue).

Figure 7.5 gives the transition intervals and colors of some selected pH indicators. Because the application of theses indicators in the context of the biochemical protocol is not volumetric analysis, the concentration of stock solution is mostly 0.1% (w/v) in ethanol or propanol, and the final dilution is 100-fold lower.

For application of the indicators in volumetric analysis and of redox and fluorescent indicators, check specialized literature.

References

Covington AK (2001) In: Lide DR (ed.) CRC Handbook of chemistry and physics, 82nd ed. CRC Press, Boca Raton, p 8–16

7.4 Buffer Recipes

The selection of buffer depends on several factors:

- The first aspect is the pH range that shall be buffered. The buffer capacity has to be large enough to catch all H^+ and OH^- ions liberated during the experiment; therefore, the pK of the

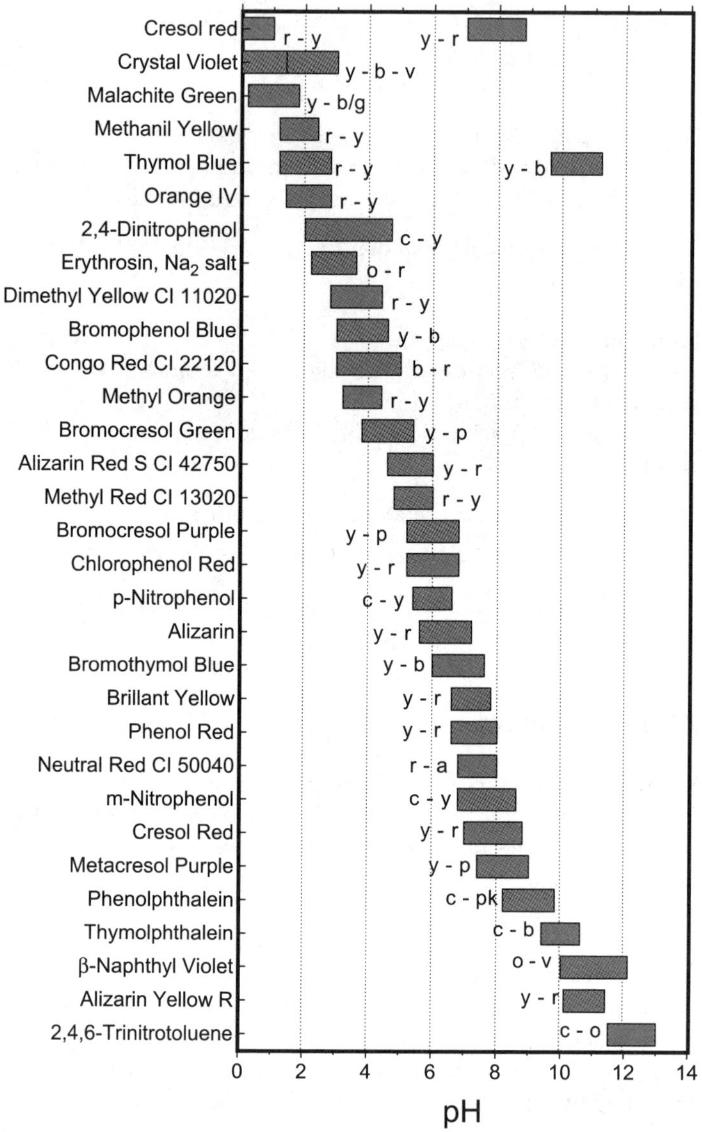

Fig. 7.5. Transition interval of pH indicators. *A* amber, *b* blue, *c* colorless, *g* green, *o* orange, *pk* pink, *p* purple, *r* red, *v* violet, *y* yellow

buffer should be near the pH which is to stabilize, and the concentration of the buffering substance should not be too low. For instance, a 10-mM Tris buffer pH 7.4 is mostly ineffective.

– A special problem is the dilution of buffer stock solutions. Especially if diluted by orders of magnitude, and if neutral salts, such as sodium chloride or potassium chloride, are added, and if the diluted buffer is used at temperatures far of its calibration, the pH of the buffer, as well as its capacity, changes dramati-

cally. It is recommended to re-equilibrate the buffer after dilution and/or mixing with salt- and protein-containing solutions. Re-equlibrate the buffer after changing the temperature, too.
- If buffer stock solutions are used, adjust them in a way that the working dilution has the right pH (pH mostly increases when buffers are diluted).
- Some buffer substances interfere with enzymes. They act as competitive inhibitors, inhibit as products, or withdraw essential ligands by chelate formation.
- Since some buffers contain reactive groups as primary amino groups, they do not suit in covalent coupling of proteins.
- Monitoring of protein content is often disturbed by UV absorption of buffers or buffer constituents.
- Some buffers have excellent characteristics with respect to biologic compatibility and/or buffering features, but they are so expensive that large-scale use is nearly impossible.

For an idea of buffering ranges, Fig. 7.6 shows these ranges for some commonly used substances.

7.4.1 Commonly Used Buffers

- *Acetate buffer, pH 4.0, $I = 0.1$*
  Mix 14.95 ml glacial acetic acid or 261 ml 1 M acetic acid with 7.875 g anhydrous sodium acetate or 96 ml 1 M sodium acetate and fill up to 1000 ml.
- *Barbital acetate buffer, pH 8.6, $I = 0.05$*
  Dissolve 4.88 g barbital-sodium (Veronal, sodium diethylbarbiturate) and 3.23 g sodium acetate trihydrate in about 800 ml ddH_2O. Adjust pH to 8.6 with 0.1 N HCl (consumption about 30 ml), then fill up to 1000 ml.
- *Barbital acetate buffer, pH 8.4, $I = 0.1$*
  Dissolve 8.14 g barbital-sodium (Veronal, sodium diethylbarbiturate) and 6.48 g sodium acetate trihydrate in about 800 ml ddH_2O. Adjust pH to 8.4 with 0.1 N HCl (consumption about 45 ml), then fill up to 1000 ml.
- *Citrate phosphate buffer, pH 5.0, 0.15 M*
  Dissolve 9.414 g citric acid and 18.155 g $Na_2HPO_4 \cdot 2H_2O$ in 900 ml ddH_2O, if necessary, correct pH by addition of diluted phosphoric acid or sodium hydroxide, then fill up to 1000 ml with ddH_2O.
  Alternatively, prepare the buffer by mixing 490 ml 0.1 M citric acid with 510 ml 0.2 M di-sodium hydrogenphosphate.
- *Glycine-HCl buffer, pH 2.5 or 2.8, 0.1 M*
  This is the most suitable buffer for elution in immunoaffinity chromatography.
  Dissolve 7.50 g glycine (aminoacetic acid, glycocoll) in 500 ml ddH_2O. Adjust the wanted pH by using 0.1 N hydrochloride acid, then fill up to 1000 ml.

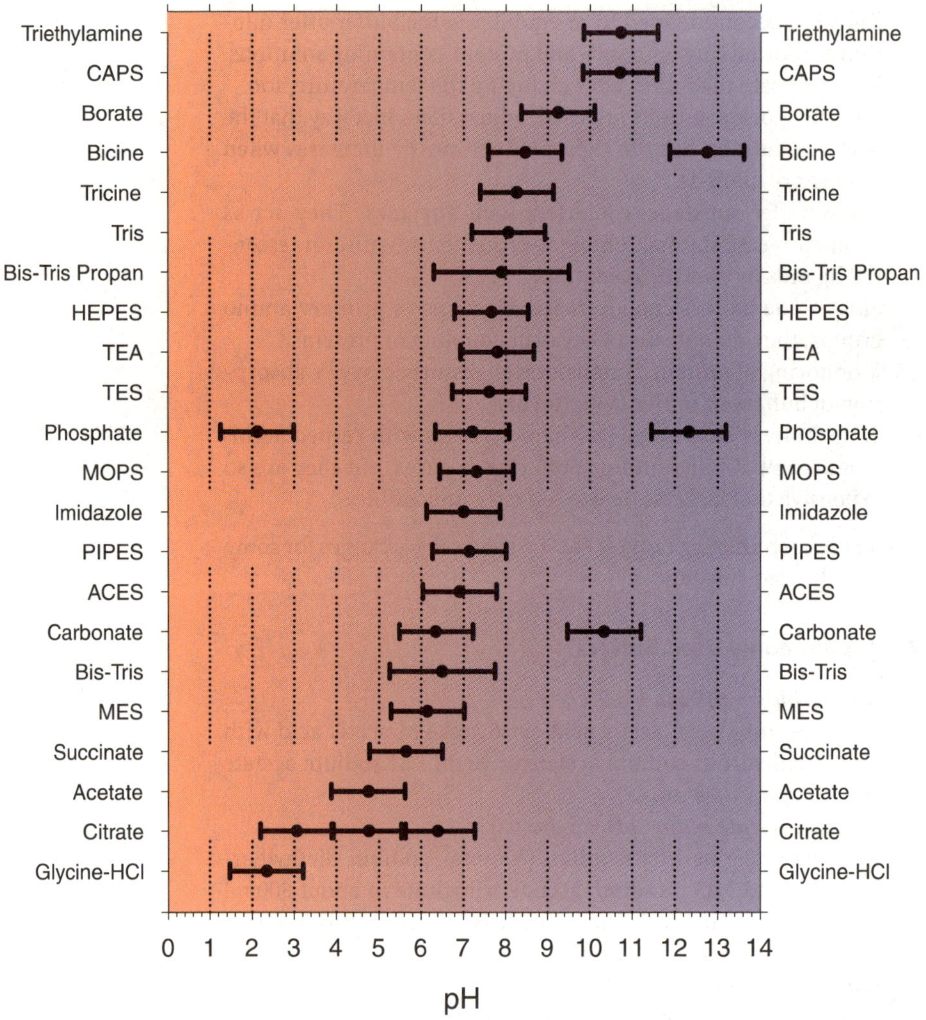

Fig. 7.6. Effective ranges of selected buffers

– *Phosphate buffer, pH 6.0–7.7, 0.01–1.0 M*
Table 7.4 gives the molar fraction x of the two basic salts of the buffer system $H_2PO_4^-/HPO_4^{--}$ for different pH and molarities M. The weighted portion g in grams of the salts Me_2HPO_4 and MeH_2PO_4 (Me: sodium or potassium) per 1000 ml buffer is calculated by

$$g\,Me_2HPO_4 = M_{r\,(Me_2HPO_4)} \cdot M \cdot x$$
$$g\,MeH_2PO_4 = M_{r\,(MeH_2PO_4)} \cdot M \cdot (1 - x)$$

The precision of the pH formed after complete dissolution is sufficient for most purposes. With respect to buffer capacity it is irrelevant if sodium or potassium salts are used.

Table 7.4. Phosphate buffer of different concentrations with pH between 6.0 and 7.7 (20 °C)

Molar fraction x of Me_2HPO_4 at total buffer molarity M										
pH	0.01	0.05	0.1	0.2	0.3	0.4	0.5	0.6	0.8	1.0 M
6.0	0.081	0.108	0.132	0.163	0.185	0.203	0.219	0.236	0.259	0.277
6.1	0.100	0.132	0.160	0.195	0.220	0.239	0.256	0.273	0.295	0.312
6.2	0.122	0.161	0.192	0.232	0.261	0.281	0.298	0.312	0.333	0.349
6.3	0.150	0.197	0.323	0.276	0.305	0.326	0.341	0.354	0.372	0.386
6.4	0.183	0.238	0.278	0.325	0.353	0.373	0.385	0.398	0.414	0.474
6.5	0.222	0.283	0.328	0.376	0.403	0.421	0.435	0.444	0.458	0.466
6.6	0.266	0.334	0.381	0.429	0.457	0.473	0.484	0.493	0.503	0.508
6.7	0.315	0.390	0.438	0.486	0.511	0.526	0.535	0.543	0.549	0.551
6.8	0.369	0.450	0.497	0.543	0.565	0.578	0.586	0.590	0.594	0.594
6.9	0.425	0.508	0.557	0.598	0.617	0.629	0.634	0.637	0.638	0.636
7.0	0.484	0.566	0.615	0.651	0.669	0.677	0.681	0.683	0.681	0.676
7.1	0.544	0.623	0.668	0.701	0.716	0.722	0.724	0.725	0.721	0.715
7.2	0.604	0.678	0.717	0.747	0.758	0.764	0.763	0.762	0.758	0.751
7.3	0.656	0.727	0.762	0.785	0.796	0.801	0.800	0.797	0.790	0.784
7.4	0.710	0.770	0.802	0.822	0.829	0.832	0.832	0.828	0.821	0.814
7.5	0.756	0.810	0.837	0.854	0.860	0.860	0.859	0.855	0.848	0.840
7.6	0.796	0.844	0.866	0.880	0.883	0.884	0.883	0.879	0.872	0.864
7.7	0.831	0.873	0.890	0.902	0.905	0.905	0.904	0.901	0.894	0.885

- *PBS (phosphate-buffered saline, isotonic (physiologic) phosphate buffer) I = 0.154, pH 7.4*
  Dissolve 7.19 g NaCl, 1.56 g KH_2PO_4, 6.74 g anhydrous Na_2HPO_4, 8.45 g $Na_2HPO_4{\cdot}2H_2O$, and 17.01 g $Na_2HPO_4{\cdot}12H_2O$ in 1000 ml ddH_2O. The final concentrations are 123 mM sodium chloride, 11.5 mM potassium phosphate, and 47.5 mM sodium phosphate.
  To avoid precipitates, PBS stock solutions should be fivefold concentrated at most.
- *PBS fivefold (Stock solution)*
  Dissolve 35.94 g NaCl, 7.82 g KH_2PO_4, 33.72 g Na_2HPO_4 (anhydrous), 42.27 g $Na_2HPO_4{\cdot}2H_2O$, and 85.06 g $Na_2HPO_4{\cdot}12H_2O$ in 1000 ml ddH_2O. To accelerate dissolution, warm the mixture to about 60 °C.
  If the stock solution is stored frozen or in a refrigerator, phosphates will crystallize. Take care to re-dissolve all precipitates before dilution to working solution.
  To protect from microbial contamination, add up to 0.02% sodium azide (w/v), 0.02% thimerosal or another biocide, and filter through a 0.2-µm filter.

- *SSC (sodium chloride-sodium citrate buffer)*
  Dissolve 8.77 g NaCl (150 mMoles) and 2.88 g citric acid (15 mMoles) in 800 ml ddH_2O. Adjust pH to 7.0 with 0.5 N sodium hydroxide, then fill up to 1000 ml.

10 × SSC

10 × SSC means the tenfold concentrated buffer. If 10 × SSC was adjusted to pH 7.0, after dilution to working concentration the pH is different to 7.0, but this difference is negligible in most cases.

- *TBS (Tris-bufferd physiologic saline)*
  Dissolve 4.25 g Tris (35 mMoles), 7.6 g NaCl (130 mMoles), 0.47 g KCl (2.5 mMoles), and 0.2 g NaN_3 (should be substituted by other biocides, because azide is an inhibitor for iron-containing enzymes) in 800 ml ddH_2O. Adjust the wanted pH between 7.6 and 8.0 with 1 N hydrochloric acid, then fill up to 1000 ml.
  TBS should not be prepared with pH below 7.4, because the buffer capacity decreases considerably.

TBS stock

TBS may be made as tenfold stock and stored frozen in aliquots. Sometimes you will find the following composition for TBS: 12.11 g Tris (0.1 Mole), 2.05 g NaCl (35 mMoles), 0.75 g glycine (10 mMoles), 0.2 g NaN_3 or Thimerosal per 1000 ml, pH 7.8–8.4.

- *TE-Buffer*
  Dissolve 1.21 g Tris (10 mMoles) and 0.372 g EDTA, disodium salt dihydrate (1 mMole EDTA) in 900 ml ddH_2O, adjust pH 8.0 with hydrochloric acid and fill up to 1000 ml. Aliquot the buffer, autoclave at 121 °C and store at 4–8 °C.

TE stock

A tenfold stock of TE buffer is possible. Check pH after dilution to working concentration, because the pK of Tris has a relatively large temperature and concentration dependency.

7.4.2 Buffers and Media for Tissue and Cell Culture and Organ Perfusion

Most of the buffers are available as pre-mixed solids or ready-to-use solutions, but preparation of these buffers makes it possible to modify the composition.

The amount of components is given in milligrams per liter (mMoles/l in parenthesis). If optically active substances are indicated, use the naturally occurring isomer. Because the addition of antibiotics depends on the intended purpose, they are not listed.

Buffers containing sodium (hydrogen) carbonate are not allowed to be autoclaved. These buffers have to be sterile filtered through a 0.2-µm filter.

- *Basal Medium Eagle with Earle's Balanced Salts* (BME modified)
  $CaCl_2$ 200 (1.80), KCl 400 (5.36), $MgSO_4 \cdot 7H_2O$ 100 (0.41), NaCl 6800 (116.44), $NaHCO_3$ 2200 (26.16), $NaH_2PO_4 \cdot H_2O$ 125 (0.80); Glucose 1000 (5.55), phenol red 10;

Arginine hydrochloride 21 (0.10), cystine 12 (0.05), glutamine 292 (2.00), histidine hydrochloride 10 (0.05), isoleucine 26 (0.20), leucine 26 (0.20), lysine hydrochloride 36.5 (0.20), methionine 7.5 (0.05), phenylalanine 16.5 (0.10), threonine 24 (0.20), tryptophane 4 (0.02), tyrosine 18 (0.10), valine 23.5 (0.20);
Biotin 1 (0.004), choline chloride 1 (0.007), folic acid 1 (0.002), i-inositol 1.8 (0.01), niacinamide 1 (0.008), calcium pantothenate 1 (0.002), pyridoxal 1 (0.005), riboflavin 0.1 (0.0003), thiamine 1 (0.003).

- *Basal Medium Eagle According to Hanks* (BME Hanks)
  $CaCl_2$ 140 (1.26), KCl 400 (5.36), KH_2PO_4 60 (0.44), $MgCl_2 \cdot 6H_2O$ 100 (0.49), $MgSO_4 \cdot 7H_2O$ 100 (0.41), NaCl 8000 (136.99), $NaHCO_3$ 350 (4.17), $Na_2HPO_4 \cdot 2H_2O$ 60 (0.34);
  Glucose 1000 (5.55), phenol red 10;
  Arginine hydrochloride 21 (0.10), cystine 12 (0.05), glutamine 292 (2.00), histidine hydrochloride 10 (0.05), isoleucine 26 (0.20), leucine 26 (0.20), lysine hydrochloride 36.5 (0.20), methionine 7.5 (0.05), phenylalanine 16.5 (0.10), threonine 24 (0.20), tryptophane 4 (0.02), tyrosine 18 (0.10), valine 23.5 (0.20);
  Biotin 1 (0.004), choline chloride 1 (0.007), folic acid 1 (0.002), i-inositol 1.8 (0.01), niacinamide 1 (0.008), calcium pantothenate 1 (0.002), pyridoxal 1 (0.005), riboflavin 0.1 (0.0003), thiamin hydrochloride 1 (0.003).
- *Dulbecco's Minimal Essential Medium* (Dulbecco's Modified Eagle's Medium, MEM Dulbecco, DMEM)
  $CaCl_2$ 200 (1.80), $Fe(NO_3)_3 \cdot 9H_2O$ 0.1 (0.0002), KCl 400 (5.36), $MgSO_4 \cdot 7H_2O$ 200 (0.81), NaCl 6400 (109.59), $NaHCO_3$ 3700 (44.05), $NaH_2PO_4 \cdot H_2O$ 125 (0.79);
  Glucose 4500 (24.97), sodium pyruvate 110 (1.00), phenol red 10;
  Arginine hydrochloride 84 (0.40), cysteine 48 (0.40), glutamine 580 (3.97), glycine 30 (0.40), histidine hydrochloride 40 (0.20), isoleucine 105 (0.80), leucine 105 (0.80), lysine hydrochloride 146 (0.76), methionine 30 (0.20), phenylalanine 66 (0.40), serine 42 (0.40), threonine 95 (0.80), tryptophane 16 (0.08), tyrosine72 (0.40), valine 94 (0.80);
  Choline chloride 4 (0.03), folic acid 4 (0.009), i-inositol 7.2 (0.04), niacinamide 4 (0.03), calcium pantothenate 4 (0.008), pyridoxal 4 (0.02), riboflavin 0.4 (0.001), thiamine hydrochloride 4 (0.01).
- *RPMI 1640 (modified)*
  $Ca(NO_3)_2 \cdot 4H_2O$ 100 (0.42), KCl 400 (5.36), $MgSO_4 \cdot 7H_2O$ 100 (0.41), NaCl 6000 (102.74), $NaHCO_3$ 2000 (23.81), Na_2HPO_4 800 (5.64);
  Glucose 2000 (11.10), phenol red 5;

Arginine hydrochloride 242 (1.15), asparagine 50 (0.38), aspartic acid 20 (0.15), cystine 50 (0.21), glutamine 300 (2.05), glutamic acid 20 (0.14), glutathione 1 (0.003), glycine 10 (0.13), histidine hydrochloride 18 (0.09), hydroxyproline 20 (0.15), isoleucine 50 (0.38), leucine 50 (0.38), lysine hydrochloride 40 (0.22), methionine 15 (0.10), phenylalanine 15 (0.09), proline 20 (0.17), serine 30 (0.29), threonine 20 (0.17), tryptophane 5 (0.02), tyrosine 20 (0.11), valine 20 (0.17);
p-Aminobenzoic acid 1 (0.007), biotin 0.2 (0.0008), choline chloride 3 (0.02), folic acid 1 (0.002), i-inositol 35 (0.19), niacinamid 1 (0.008), calcium pantothenate 0.25 (0.0005), pyridoxine hydrochloride 1 (0.005), riboflavin 0.2 (0.0005), thiamine hydrochloride 1 (0.003), vitamin B_{12} (cyanocobalamin) 0.005.

- *RPMI 1640 HAT*
  Supplement 100 ml RPMI 1640 with 1 ml HT stock and 1 ml A stock.
  HT Stock solution: Make slurry of 408 mg hypoxanthine in 100 ml ddH_2O, then add 1 M sodium hydroxide until all solid is dissolved. Dissolve 114 mg thymidine in 100 ml ddH_2O. Combine both solutions, fill up to 300 ml, and adjust pH with acetic acid to 10.0. Filter through a 0.2-μm filter and store in aliquots at −20 °C.
  A Stock solution: Vortex 45.4 mg aminopterine (4-aminopteroyl glutamic acid, sodium salt) in 5 ml ddH_2O and dissolve the solid by dropwise addition of 1 M NaOH. Fill up to 100 ml and adjust pH to 7.5. Filter through a 0.2-μm filter and store at −20 °C.
- *Krebs-Henseleit-Ringer Buffer*
  $CaCl_2$ 280 (2.53), KCl 354 (4.74), KH_2PO_4 162 (1.19), $MgSO_4 \cdot 7H_2O$ 292 (1.19), NaCl 6923 (118.54), $NaHCO_3$ 2100 (25.00).
- *Tyrode Solution*
  $CaCl_2$ 200 (1.80), KCl 200 (2.68), NaCl 8000 (136.99), $NaHCO_3$ 1000 (11.90), $Na_2HPO4 \cdot 2H_2O$ 60 (0.34), Glucose 1000 (5.55).

7.4.3 pH Calibration Buffers

A potassium hydrogentartrate solution, saturated at 25 °C
B 50 mM potassium hydrogenphthalate: 1.0212 g $C_8H_5KO_4$ per 100 ml
C 25 mM phosphate buffer: 0.3403 g KH_2PO_4; 0.4450 g $Na_2HPO_4 \cdot 2H_2O$ per 100 ml
D 8.7 mM phosphate buffer: 0.1183 g KH_2PO_4; 0.5416 g $Na_2HPO_4 \cdot 2H_2O$ per 100 ml
E 10 mM sodium tetraborate (borax): 0.2012 g $Na_2B_4O_7$ per 100 ml

The pH values derived by the standard buffers are temperature dependent. The pH values between 0 and 60 °C are given in Table 7.5, which also indicates the change of pH at 1:1 dilution ($\Delta pH_{1:1}$).

Table 7.5. pH of calibration buffers between 0 and 60 °C

Buffer Temperature (°C)	A	B	C pH	D	E
0	–	4.00	6.98	7.53	9.46
5	–	4.00	6.95	7.5	9.395
10	–	4.00	6.92	7.47	9.33
15	–	4.00	6.9	7.45	9.27
20	–	4.00	6.88	7.43	9.23
25	3.56	4.01	6.865	7.41	9.18
30	3.55	4.015	6.85	7.4	9.14
35	3.55	4.025	6.855	7.39	9.1
38	3.55	4.03	6.84	7.38	9.08
40	3.55	4.035	6.84	7.38	9.07
45	3.55	4.05	6.845	7.37	9.04
50	3.55	4.06	6.83	7.37	9.01
55	3.555	4.075	6.83	–	8.98
60	3.56	4.09	6.835	–	8.96
Δ $pH_{1:1}$	+0.052		+0.08	+0.08	+0.01

References

Beynon RJ, Easterby JS (1996) Buffer solutions. IRL Press, Oxford, p 83
Covington AK (2001) In: Lide DR (ed.) CRC Handbook of chemistry and physics, 82nd ed. CRC Press, Boca Raton, p 8-16
Stoll VS, Blanchard JS (1990) Meth Enzymol 182:24

7.4.4 Volatile Buffers

Sometimes it is helpful to use buffers that are removed during lyophilization. Table 7.6 presents some of these volatile buffers. It should be mentioned that a part of buffer components are bound as counter ions to the charged residues of proteins, i.e., the lyophilized sample is not a completely freed form buffer compounds.

Table 7.6. Volatile buffers

Buffer	pH range
Pyridine-formic acid	2.3 – 3.5
Trimethylamine-formic acid	3.0 – 5.0
Pyridine-acetic acid	3.0 – 6.0
Trimethylamine-hydrochloric acid	6.8 – 8.8
Ammonia-formic acid	7.0 – 8.5
Triethylamine-carbon dioxide	7.0 – 12.0
Ammonium carbonate-ammonia	8.0 – 9.5
Ammonia-hydrochloric acid	8.5 – 10.0

8 Tables

8.1 Concentration Units

Table 8.1. Concentration units

Concentration unit	Symbol	Definition	Description
Molarity	M	$\text{Mol} \cdot l^{-1}$	Moles per liter solution
Molality	–	$\text{Mol} \cdot kg^{-1}$	Moles per kilogram solvent
Normality	N	$\text{Mol} \cdot l^{-1} \cdot n^{-1}$	Moles per liter solution and equivalent
Normality	N	$\text{Val} \cdot l^{-1}$	Vales per liter solution
Percentage weight per volume	% (w/v)		Grams per 100 ml solution
Percentage weight per weight	% (w/w)		Grams per 100 g solution
Percentage volume per volume	% (v/v)		Milliliters per 100 ml solution
Parts per million parts	ppm	10^{-4}%	e.g., ng/mg
Parts per billion parts	ppb	10^{-7}%	e.g., pg/mg

8.2 Conversion Factors for SI Units

Table 8.2. Selected SI units

Quantity	Unit	Symbol	Definition
Activity of a radionuclide	Becquerel	Bq	$1\ \mathrm{Bq} = 1\ \mathrm{s}^{-1}$
Amount of substance	Mole	mol	(basic SI unit)
Electric charge	Coulomb	C	$1\ \mathrm{C} = 1\ \mathrm{s} \times \mathrm{A}$
Electric current	Ampere	A	(basic SI unit)
Energy	Joule	J	$1\ \mathrm{J} = 1\ \mathrm{m}^2 \times \mathrm{kg} \times \mathrm{s}^{-2}$
Enzyme activity	Katal	kat	$1\ \mathrm{kat} = 1\ \mathrm{mol} \times \mathrm{s}^{-1}$
Force	Newton	N	$1\ \mathrm{N} = 1\ \mathrm{m} \times \mathrm{kg} \times \mathrm{s}^{-2}$
Frequency	Hertz	Hz	$1\ \mathrm{Hz} = 1\ \mathrm{s}^{-1}$
Length	Meter	m	(basic SI unit)
Mass	Kilogram	kg	(basic SI unit)
Power	Watt	W	$1\ \mathrm{W} = 1\ \mathrm{m}^2 \times \mathrm{kg} \times \mathrm{s}^{-3}$
Pressure	Pascal	Pa	$1\ \mathrm{Pa} = 1\ \mathrm{m}^{-1} \times \mathrm{kg} \times \mathrm{s}^{-2}$
Temperature	Kelvin	K	(basic SI unit)
Time	Second	s	(basic SI unit)

Table 8.3. SI prefixes

Prefix	Symbol	Factor	Prefix	Symbol	Factor
yotta	Y	10^{24}	Deci	d	10^{-1}
zetta	Z	10^{21}	Centi	c	10^{-2}
exa	E	10^{18}	Milli	m	10^{-3}
peta	P	10^{15}	Micro	µ	10^{-6}
tera	T	10^{12}	Nano	n	10^{-9}
giga	G	10^{9}	Pico	p	10^{-12}
mega	M	10^{6}	Femto	f	10^{-15}
kilo	k	10^{3}	Atto	a	10^{-18}
hecto	h	10^{2}	Zepto	z	10^{-21}
deca	da	10^{1}	Yocto	y	10^{-24}

Table 8.4. Fundamental physical constants

Constant	Value	Unit	Symbol
Avogadro constant	6.022×10^{23}	mol^{-1}	N_A; L
Elementary charge	1.602176×10^{-19}	C	e
Molar gas constant	8.3145	$J \times mol^{-1} \times K^{-1}$	R
Newtonian constant of gravitation	6.673×10^{-11}	$m^3 \times kg^{-1} \times s^{-2}$	G

Table 8.5. Conversion factors into SI units

Unit	Symbol	Conversion to SI unit
Ångström unit	Å	1 Å = 0.1 nm
Atmospheres (physical)	atm	1 atm = 101.325 kPa
Atmospheres (technical)	at	1 at = 98.066 kPa
Bar	bar	1 bar = 100 kPa
Calorie	cal	1 cal = 4.1868 J
Centimeter of water	cm H_2O	1 cm H_2O = 98.07 Pa
Curie	Ci	1 Ci = 37 GBq
Degree centigrade (°Celsius)	°C	x °C = K – 273.15
Degree Fahrenheit	°F	1 °F = 0.555 °C x °C = (y °F – 32)/1.8 x K = (459.67 + y °F)/1.8
Desintegrations per minute	dpm	1 dpm = 0.0167 Bq
Electron volt	eV	1 eV = 1.60202×10^{-19} J
Erg	erg	1 erg = 10^{-7} J
Foot	ft	1 ft = 0.3048 m
Gamma	γ	1 γ = 1 μg
Grain	gr	1 gr = 64.799 mg
Inch	in	1 in = 25.4 mm
International Unit (enzyme activity)	IU	1 IU = 16.67 nkat
Kilopond	kp	1 kp = 9.80665 N
Kilopond/cm^2	kp/cm^2	1 kp/cm^2 = 98.066×10^3 Pa
Millimeter of mercury	mm Hg	1 mmHg = 133.32 Pa
Pound	lb	1 lb = 0.4536 kg
Pound per square inch	psi	1 psi = 6.8948×10^3 Pa
Torr	Torr	1 Torr = 133.32 Pa

8.3 Data of Frequently Used Substances

Amino acids: see Table 8.13;
Nucleotides: see Table 8.15;
Buffers: see Table 7.2
Organic solvents: see Table 8.10

Table 8.6. Relative atomic mass M_r of naturally occurring elements (basis $^{12}C = 12.00$)

Name	Symbol	Atomic weight	Name	Symbol	Atomic weight
Aluminum	Al	26.98	Neodymium	Nd	114.24
Antimony	Sb	121.75	Neon	Ne	20.18
Argon	Ar	39.95	Nickel	Ni	58.71
Arsenic	As	74.92	Niobium	Nb	92.91
Barium	Ba	137.34	Nitrogen	N	14.01
Beryllium	Be	9.01	Osmium	Os	190.2
Bismuth	Bi	208.98	Oxygen	O	15.999
Boron	B	10.81	Palladium	Pd	106.42
Bromine	Br	79.91	Phosphorus	P	30.97
Cadmium	Cd	112.41	Platinum	Pt	195.09
Calcium	Ca	40.08	Potassium	K	39.098
Carbon	C	12.01	Preseodymium	Pr	140.91
Cerium	Ce	140.12	Radium	Ra	226
Cesium	Cs	132.91	Radon	Rn	222
Chlorine	Cl	35.45	Rhenium	Re	186.21
Chromium	Cr	51.996	Rhodium	Rh	102.91
Cobalt	Co	58.93	Rubidium	Rb	85.47
Copper	Cu	63.55	Ruthenium	Ru	101.07
Dysprosium	Dy	162.5	Samarium	Sm	150.35
Erbium	Er	167.26	Scandium	Sc	44.96
Europium	Eu	151.96	Selenium	Se	78.96
Fluorine	F	18.998	Silicon	Si	28.09
Gadolinium	Gd	157.25	Silver	Ag	107.87
Gallium	Ga	69.72	Sodium	Na	22.99
Germanium	Ge	72.59	Strontium	Sr	87.62
Gold	Au	196.97	Sulfur	S	32.06
Hafnium	Hf	178.49	Tantalum	Ta	180.95
Helium	He	4.00	Tellurium	Te	127.6
Holmium	Ho	164.93	Terbium	Tb	158.92
Hydrogen	H	1.01	Thallium	Tl	204.37
Indium	In	114.82	Thorium	Th	232.04
Iodine	I	126.90	Thulium	Tm	168.93
Iridium	Ir	192.22	Tin	Sn	118.69
Iron	Fe	55.85	Titanium	Ti	47.87

Table 8.6. (continued)

Name	Symbol	Atomic weight	Name	Symbol	Atomic weight
Krypton	Kr	83.80	Tungsten	W	183.85
Lanthanum	La	138.91	Uranium	U	238.03
Lead	Pb	207.2	Vanadium	V	50.94
Lithium	Li	6.94	Xenon	Xe	131.29
Luthetium	Lu	174.97	Ytterbium	Yb	173.04
Magnesium	Mg	24.31	Yttrium	Y	88.91
Manganese	Mn	54.94	Zinc	Zn	65.37
Mercury	Hg	200.59	Zirconium	Zr	91.22
Molybdenum	Mo	95.94			

Table 8.7. Relative molecular mass M_r of multiples of atoms and residues

Residue	M_r	Residue	M_r	Residue	M_r
C_1	12.01	H_1	1.01	OCH_3	31.03
C_2	24.02	H_2	2.02	$(OCH_3)_2$	62.07
C_3	36.03	H_3	3.02	$(OCH_3)_3$	93.10
C_4	48.04	H_4	4.03	$(OCH_3)_4$	124.14
C_5	60.05	H_5	5.04	$(OCH_3)_5$	155.17
C_6	72.06	H_6	6.05		
C_7	84.07	H_7	7.06	OCH_2CH_3	45.06
C_8	96.08	H_8	8.06	$(OCH_2CH_3)_2$	90.12
C_9	108.09	H_9	9.07	$(OCH_2CH_3)_3$	135.18
C_{10}	120.10	H_{10}	10.08		
				CO_2CH_3	59.04
CH_3	15.03	H_2O	18.02	$(CO_2CH_3)_2$	118.09
$(CH_3)_2$	30.07	$(H_2O)_{0.5}$	9.01		
$(CH_3)_3$	45.10	$(H_2O)_2$	36.03	$COOH$	45.02
$(CH_3)_4$	60.14	$(H_2O)_3$	54.05	$(COOH)_2$	90.04
$(CH_3)_5$	75.17	$(H_2O)_4$	72.06	$(COOH)_3$	135.05
$(CH_3)_6$	90.18	$(H_2O)_5$	90.08		
		$(H_2O)_6$	108.11	$CHOH$	30.03
C_2H_5	29.06			$(CHOH)_2$	60.05
$(C_2H_5)_2$	58.12	NH_2	16.03	$(CHOH)_3$	90.08
$(C_2H_5)_3$	87.18	$(NH_2)_2$	32.05	$(CHOH)_4$	120.10
$(C_2H_5)_4$	116.24	$(NH_2)_3$	48.08	$(CHOH)_5$	150.13
(C_2H_5)	145.30	$(NH_2)_4$	92.11	$(CHOH)_6$	180.16

Table 8.8. Relative molecular mass M_r of inorganic compounds

Formula	M_r	Formula	M_r
$AgNO_3$	169.87	KSCN	97.18
$AuHCl_4$	339.79	$MgCl_2 \times 6H_2O$	203.30
$BaCl_2$	208.25	$Mg(acetate)_2 \times H_2O$	214.46
$CaCl_2 \times 6H_2O$	219.08	$MnCl_2 \times H_2O$	197.91
$Cu(acetate)_2 \times H_2O$	199.65	NH_4OH	35.05
$CuSO_4 \times 5H_2O$	249.68	$(NH_4)_2SO_4$	132.13
$FeCl_3$	162.21	$Na_2B_4O_7 \times 10H_2O$	381.37
$FeSO_4 \times 7H_2O$	278.01	NaCl	58.44
HCl	36.46	$NaClO_4$	122.44
$HClO_4$	100.46	Na_2CO_3	105.99
HNO_3	63.01	$Na_2CO_3 \times 10H_2O$	286.14
H_2BO_3 (boric acid)	61.83	$NaHCO_3$	84.01
H_2O	18.02	NaH_2PO_4	119.98
H_3PO_4	98.00	$NaH_2PO_4 \times 2H_2O$	156.01
H_2SO_4	98.08	Na_2HPO_4	141.96
KCl	74.55	$Na_2HPO_4 \times 2H_2O$	177.99
KH_2PO_4	136.09	$Na_2HPO_4 \times 12H_2O$	358.14
$KH_2PO_4 \times 3H_2O$	228.23	NaI	149.89
K_2HPO_4	174.18	$NaIO_4$	213.89
KI	166.00	NaOH	40.00
KOH	56.11	NaSCN	81.07

Table 8.9. Organic compounds

Name	M_r	Specific gravity[a]
Acetic acid	60.05	1.049
N-Acetylglucosamine (GlcNAc)	221.21	
D-Biotin	244.30	
Butyric acid	88.11	0.964
Chloralhydrate	165.40	
Cholic acid	408.56	
Citric acid, trisodium salt, dihydrate	294.10	
Citric acid	192.12	
Deoxycholic acid, sodium salt (NaDOC)	414.56	
DTE (Dithioerythritol, CLELAND's Reagent)	154.25	
DTT (Dithiothreitol, CLELAND's Reagent)	154.25	
EAC (ε-aminocaproic acid, 6-aminohexanic acid)	131.18	
EDTA (Ethylendiaminotetraacetic acid, disodium salt)	372.24	
EGTA (Ethyleneglycol-bis(2-aminoethyl-ether)-N,N,N′,N′-tetraacetic acid)	380.40	
Ethylenediamine	60.10	0.899
Ethyleneglycol	62.07	1.113
Formaldehyde	30.03	
Formic acid	46.03	1.220
Glucose	180.16	
Glutaraldehyde	100.12	
Glycerol	92.09	1.261
Guanidinium hydrochloride	95.53	
Hexamethylenediamine (1,6-Diaminohexane)	116.21	
Iodoacetamide	184.96	
2-Mercaptoethanol	78.13	1.114
Methyl-α-D-mannopyranoside	194.18	
PMSF (Phenylmethylsulfonylfluoride)	174.19	
n-Propanol	60.10	0.804
SDS (Dodecylsulfat, sodium salt)	288.38	
Sucrose	342.30	
Sulfosalicylic acid dihydrate	254.21	
TCA (Trichloroacetic acid)	163.39	
Thimerosal (2-(Ethylmercurimercapto)-benzoic acid, sodium salt; Merthiolat)	404.81	
Thioglycolic acid (Mercaptoacetic acid)	92.12	1.325
Thioglycolic acid, sodium salt	114.10	
Tris (Tris(hydroxymethyl)aminomethane)	121.14	
Urea	60.06	

[a] For liquids, related to 20 °C (in g/cm^3)

Table 8.10. Organic solvents (molar mass, specific gravity, boiling point, melting point, flash point)

Solvent	M_r	Density (g/cm^3)	Bp	Mp (°C)	Flash point
Acetic acid (glacial acetic acid)	60.06	1.05	118.0	17.0	40
Acetone	58.08	0.79	56.0	−95.0	−20
Acetonitrile	41.05	0.78	82.0	−46.0	5
Benzen	78.11	0.88	80.0	5.5	−11
tert-Butanol	74.12	0.77	82.0	24.0	11
Carbon tetrachloride (Tetra)	154.82	1.60	77.0	−23.0	
Chloroform	119.38	1.47	62.0	−64.0	
Dichloromethan	84.93	1.32	39.0	−97.0	
Diethylether (anaesthesia ether)	74.12	0.71	35.0		−116
Dimethylformamide (DMF)	73.10	0.95	153.0	−61.0	59
Dimethylsulfoxide (DMSO)	78.13	1.10	189.0	18.5	95
1,4-Dioxan	88.11	1.03	101.0	12.0	12
Ethanol	46.07	0.79	79.0	−117.0	9
Ethylacetate (acetic ester)	88.10	0.90	77.0	−84.0	−4
Ethylene glycol (ethane-1,2-diol)	62.07	1.11	198.0	−13.0	111
Formamide	45.04	1.13	210.0	2.0	175
n-Hexane	86.18	0.66	69.0	−95.0	−22
Methanol	32.04	0.79	65.0	−93.0	12
n-Butanol	74.12	0.81	117.7	−90.0	35
2-Propanol (isopropanol)	60.10	0.78	82.0	−89.0	12
Pyridine	79.10	1.51	115.6	−41.8	17
Tetrahydrofuran (THF)	72.11	0.89	66.0	−108.0	−20
Toluen	92.14	0.87	111.0	−95.0	4
Triethanolamine	149.19	1.12	360.0	22.0	179

8.4 Protein Data

Table 8.11

References

Fasman GD (ed.) (1992) Practical CRC handbook of biochemistry and molecular biochemistry. CRC Press, Boca Raton, p 196
Gill SC, von Hippel PH (1989) Anal Biochem 182:319
Kirschenbaum DM (1978) Anal Biochem 87:223

Table 8.11. Molecular mass, isoelectric point, and absorption coefficient of proteins)

Protein	M_r (kD) Holoprotein	Subunit	pI	A_λ c = 1 mg/ml	λ (nm)	Species
Aldolase	160	40	5.2	0.74	280	Rabbit
Aldolase	80			1.06	280	Yeast
Alkaline phosphatase	140	69	4.4	1.00	280	Cattle
Alkaline phosphatase	80			0.77	280	*Escherichia coli*
Alkohol dehydrogenase	80	41	8.7 – 9.3	0.46	280	Horse
Alcohol dehydrogenase	141	35	5.4	1.26	280	Yeast
D-Amino acid oxidase	100	50		1.60	280	Pig
α-Amylase	51.3			2.50	280	Pig
Apoferritin	443	18.5	4.1 – 4.5	0.86 – 0.97	280	Horse
Avidin	67	16.8		1.57	280	Chicken
Carbonic acid anhydrase	30		6	1.80	280	Cattle
Carboxypeptidase A	34.3			1.81	278	Cattle
Catalase	257	57.5		1.68	405	Cattle
Chymontrypsinogen	25.761			2.00	280	Cattle
Collagenase	105	57		1.47	280	*Clostridium histolyticum*
Concanavalin A	108	54; 26		1.14	280	*Canavalis ensiformis*
Cytochrom c	12.75		9.6	2.39	550	Horse
Cytochrom c	12.75			0.20	280	Pig

Table 8.11. (continued)

Protein	M_r (kD) Holoprotein	Subunit	pI	A_λ c = 1 mg/ml	λ (nm)	Species
Cytochrom c oxidase	200	100		1.74	280	Cattle
Enolase	88	41		0.90	280	Yeast
Fibrinogen	341	63.5; 56; 47	5.5	1.39	280	Man
Glucagon	3.483			2.38	278	Cattle
Glucose oxidase	186	80	4.15	1.67	280	*Aspergillus niger*
Glutamate dehydrogenase	56.1			0.95	280	Cattle
Hemocyanine (KLH)	> 2000			1.39	280	*Limulus polyphemus*
Hemocyanine (KLH)	> 2000			0.22	340	*Limulus polyphemus*
Hemoglobin	68	16.5		0.80	280	Man
Hexokinase	102	51	4.93; 5.25	0.92	280	yeast
Immunoglobulin IgA	180–500			1.34	280	Man
Immunoglobulin IgD	175		4.7–6.1	1.45	280	Man
Immunoglobulin IgE	200			1.41	280	Man
Immunoglobulin IgG	150–160	~50; ~25	5.8–7.3	1.38	280	Man
Immunoglobulin IgM	950			1.33	280	Man
Immunoglobulin IgY	190	73; 22		1.35	275	Chicken
Insulin	5.733		5.5	1.00	280	Cattle
Insulin A chain	2.33					
Insulin B chain	3.4					

Table 8.11. (continued)

Protein	M_r (kD) Holoprotein	Subunit	pI	A_λ c = 1 mg/ml	λ (nm)	Species
α-Lactalbumin	14.4			2.09	280	Cattle
Lactate dehydrogenase	134	35	8.5	1.50	280	Cattle
β-Lactoglobulin	35	17.5	5.2	0.95	280	Cattle
Luciferase	92	52				Glow worm
Lysozyme	14.388		11	2.53	280	Chicken
α2-Macroglobulin	820	180–190	5.4	0.81	280	Man
Myoglobin	16.95		6.85; 7.35			Horse
Myosin	205–215		4.8–6.2	0.52	280	Rabbit
Myosin L cahin	18			0.43	277	Rabbit
Ovalbumin	43			0.75	280	Chicken
Ovotransferrin	76.6			1.16	280	Chicken
Pepsin	33.367		2.2	1.48	280	Cattle
Pepsinogen	41		3.7	1.31	280	Pig
Peroxidase	40		7.2	1.34	275	Horse raddish
Phosphodiesterase	115					*Crotus adamanteus*
Phosphorylase b	97.4			1.32	280	Rabbit
Protein A	42			0.17	280	*Staphylococcus aureus*
Protein G	30–35					*Streptococcus C* or *G*

Table 8.11. (continued)

Protein	M_r (kD) Holoprotein	Subunit	pI	A_λ c = 1 mg/ml	λ (nm)	Species
Pyruvate kinase	237	57.2	7.8–8.6	0.54	280	Rabbit
Ribonuclease I	13.683		7.8	0.70	280	Cattle
RNA polymerase	500	39; 90; 155				*Escherichia coli*
Serum albumin	66.398		4.9	0.67	280	Cattle
Serum albumin	66.437		4.7	0.51	280	Man
β-Galactosidase	540	135		1.91	280	*Escherichia coli*
Streptavidin	60	15	5	3.40	280	*Streptomyces avidinii*
Trypsin	23.8			1.66	280	Cattle
Trypsin inhibitor	20.1			0.95	280	Soy bean
Trypsinogen	24.5			1.39	280	Cattle
Urease	483	240		0.59	280	*Canavalis ensiformis*
Uricase	125	32		1.13	276	Pig

8.5 Protease Inhibitors

In the literature there are a lot of more or less specific inhibitors of proteases. During daily work with tissue homogenates, a mixture of inhibitors with broad specificity has been proved with success. This mixture is given in Table 8.12.

The individual inhibitors are made as stock solutions, mostly 1000-fold concentrated with respect to final dilution. Add the stock solutions immediately before starting homogenization.

Inhibitor Stock Solutions

0.1 M PMSF	dissolve 87 mg into 5 ml i-propanol or ethanol; stable for 4–6 weeks when refrigerated
1 M Iodoacetamide	dissolve 925 mg in 5 ml ddH_2O; stable at RT
0.5 M EDTA	dissolve 1.86 g EDTA, disodium salt, in about 8 ml ddH_2O, adjust pH 8 with sodium hydroxide and fill up to 10.0 ml; stable at RT
Leupeptin	dissolve 2.5 mg in 5 ml ddH_2O; stable for 1 week when refrigerated
0.1 M Benzamidine	dissolve 87 mg in 5 ml ddH_2O; stable at RT
0.1 M o-Phenanthroline	dissolve 99 mg in about 4 ml ddH_2O, adjust pH 6 with hydrochloric acid, fill up to 5.0 ml; stable at RT
Pepstatin A	1 mg/ml in methanol; stable at –20 °C for several weeks

Further inhibitors (working concentrations in µg/ml in parenthesis) are:

4-(2-aminoethyl)-benzensulfonyl fluoride (100–1000; Pefabloc SC, AEBSF, use instead of PMSF or 4-amidinophenyl-methylsulfonyl fluoride (10–40; APMSF); antipain (50); aprotinin (0.06–2); bestaine (40); calpain inhibitor I (17; N-acetyl-Leu-Leu-neulorcinal); N-[N-L-3-trans-carboxiran-2-carbonyl)-L-leucyl]-4-aminobutyl guanidin (0.5–10; E-64); L-1-chloro-3-(4-tosylamido)-7-amino-2-heptanone (37 to 50; TLCK); L-1-chloro-3-(4-tosylamido)-4-phenyl-2-butanone (70 to 100; TPCK); chymostatin (6–60; N-(α-rham-

Table 8.12. Protease inhibitor mix

Inhibitor	Final concentration	Specificity
Benzamidine	0.1 mM	Trypsin and similar proteases
EDTA	5 mM	Metalloproteases
Iodacetamide	1 mM	Thiol proteases
Leupeptin	0.5 µg/ml	(broad specificity)
Pepstatin A	0.5 µg/ml	Acidic proteases
o-Phenanthroline	0.1 mM	(broad specificity)
PPSF	0.1 mM	Serine and thiol proteases

nopyranosyl hydroxyphosphinyl)-L-leucyl-L-tryptophane, Phosphoramidon); 3,4-dichloroisocoumarine (1–43); EGTA (200–500); hirudin (150–200 ATU); phosphoamidon (4–330), tissue inhibitor of metalloproteinase 2 (10; TIMP-2); trypsin inhibitor (from chicken egg white or soy bean; 10–100).

References

Zöllner H (1992) Handbook of enzyme inhibitors, 2nd ed. VCH, New York

8.6 Single-Letter Codes and Molecular Masses of Amino Acids

The single-letter code is helpful to describe longer amino acid sequences. It is derived from the three-letter code of amino acids (aa).

Checking number CN and reference number AA is used for unequivocal characterization of each amino acid sequence:

$$CN = \sum_{i=1}^{NR} AA_i \cdot i$$

Example

Sequence (IUPAC-IUB): His-Glu-Leu-Pro-Met-Ala-Thr-His-Glu
Single-letter code: HELPMATHE
CN: $441 = 1 \cdot 9 + 2 \cdot 7 + 3 \cdot 11 + 4 \cdot 15 + 5 \cdot 13 + 6 \cdot 1 + 7 \cdot 17 + 8 \cdot 9 + 9 \cdot 7$
NR: 9
M_r: 1064.19
COMP: $A_1 R_0 N_0 D_0 C_0 Q_0 E_2 G_0 H_2 I_0 L_1 K_0 M_1 F_0 P_1 S_0 T_1 W_0 Y_0 V_0$

CN: checking number; AA: reference number (cf. Table 8.13); i: position number of the respective aa within the sequence (i of the N-terminal aa is 1); NR: total number of aa within the polypeptide chain; M_r: molar mass of the polypeptide; COMP: true formula of the polypeptide (all naturally occurring aa have to be listed, aa which are no part of the chain are indicated by "0").

References

Bairoch A (1982) Biochem J 203:527

Table 8.13. Amino acids: code and molecular mass

Amino acid	Triple-letter code	Single-letter code	M_r	$M_r - H_2O$	AA
Alanine	Ala	A	89.09	71.08	1
Arginine	Arg	R	174.20	156.19	2
Asparagine	Asn	N	132.12	114.11	3
Aspartic acid	Asp	D	133.10	115.09	4
Citrulline			175.19	157.18	
Cysteine	Cys	C	121.16	103.15	5
Cystine	Cys_2		240.30	222.29	
γ-Carboxyglutamic acid	Gla		191.14	173.13	
Glutamic acid	Glu	E	147.13	129.12	7
Glutamine	Gln	Q	146.15	128.14	6
Glycine	Gly	G	75.07	57.06	8
Histidine	His	H	155.16	137.15	9
Homocysteine			135.19	117.18	
5-Hydroxylysine			162.19	144.18	
trans-4-Hydroxyproline	Hyp		131.13	113.12	
Isoleucine	Ile	I	131.17	113.16	10
Leucine	Leu	L	131.17	113.16	11
Lysine	Lys	K	146.19	128.18	12
Methionine	Met	M	149.21	131.20	13
Norleucine	Nle		131.17	113.16	
Norvaline	Nva		117.15	99.14	
Ornithine	Orn	O	132.16	114.15	
Phenylalanine	Phe	F	165.19	147.18	14
Proline	Pro	P	115.13	97.12	15
Pyroglutamic acid	pGlu, Pyr		129.12	111.11	
Selenocysteine	Sec		168.05	150.04	
Serine	Ser	S	105.09	87.08	16
Statine	Sta		175.23	157.22	
Threonine	Thr	T	119.12	101.11	17
Tryptophane	Trp	W	204.23	186.22	18
Tyrosine	Tyr	Y	181.19	163.18	19
Valine	Val	V	117.17	99.16	20
Asparagine or aspartic acid	Asx	B	132.65	114.64	21
Glutamine or glutamic acid	Glx	Z	146.64	128.63	22
Unknown amino acid	Xaa	X	144.56	126.55	23

Table 8.14. Genetic code

	U/A		C/G		A/T		G/C		
U/A	UUU AAA	Phe/F	UCU AGA	Ser/S	UAU ATA	Tyr/Y	UGU ACA	Cys/C	U/A
	UUC AAG	Phe/F	UCC AGG	Ser/S	UAC ATG	Tyr/Y	UGC ACG	Cys/C	C/G
	UUA AAT	Leu/L	UCA AGT	Ser/S	UAA ATT	Stop	UGA ACT	Sec Stop	A/T
	UUG AAC	Leu/L	UCG AGC	Ser/S	UAG ATC	Stop	UGG ACC	Trp/W	G/C
C/G	CUU GAA	Leu/L	CCU GGA	Pro/P	CAU GTA	His/H	CGU GCA	Arg/R	U/A
	CUC GAG	Leu/L	CCC GGG	Pro/P	CAC GTG	His/H	CGC GCG	Arg/R	C/G
	CUA GAT	Leu/L	CCA GGT	Pro/P	CAA GTT	Gln/Q	CGA GCT	Arg/R	A/T
	CUG GAC	Leu/L	CCG GGC	Pro/P	CAG GTC	Gln/Q	CGG GCC	Arg/R	G/C
A/T	AUU TAA	Ile/I	ACU TGA	Thr/T	AAU TTA	Asn /N	AGU TCA	Ser/S	U/A
	AUC TAG	Ile/I	ACC TGG	Thr/T	AAC TTG	Asn /N	AGC TCG	Ser/S	C/G
	AUA TAT	Ile/I	ACA TGT	Thr/T	AAA TTT	Lys/K	AGA TCT	Arg/R	A/T
	AUG TAC	Met/M Start	ACG TGC	Thr/T	AAG TTC	Lys/K	AGG TCC	Arg/R	G/C
G/C	GUU CAA	Val/V	GCU CGA	Ala/A	GAU CTA	Asp/D	GGU CCA	Gly/G	U/A
	GUC CAG	Val/V	GCC CGG	Ala/A	GAC CTG	Asp/D	GGC CCG	Gly/G	C/G
	GUA CAT	Val/V	GCA CGT	Ala/A	GAA CTT	Glu/E	GGA CCT	Gly/G	A/T
	GUG CAC	Val/V	GCG CGC	Ala/A	GAG CTC	Glu/E	GGG CCC	Gly/G	G/C

Wobble bases

M	A C
V	A C G
N	A C G T
H	A C T
R	A G
D	A G T
W	A T
B	C G T
Y	C T
S	G C
K	G T
I	Inosin
s	Thio
X	Modif

Examples for wobble bases:

ACN	Thr
MGN	Arg

mRNA	5′→ 3′
Protein	N→ C
cDNA	3′→ 5′

Coding in:

RNA

DNA

8.7 Spectroscopic Data of Nucleotides

References

Dawson RMC, Elliott DC, Elliott WH, Jones KM (1986) Data for biochemical research, 3rd ed. Clarendon Press, Oxford

Table 8.15. Nucleotides: spectroscopic data and molecular mass

Nucleotide	λ_{max} (nm)	ε_{max} $\times 10^{-3}$	ε_{260}	M_r	M_r (stable formulation)
5′-AMP	259	15.4	15.4	347.2	365.2 ($\times H_2O$)
5′-ADP	259	15.4	15.4	427.2	471.2 (Na_2 salt)
5′-ATP	259	15.4	15.4	507.2	605.2 ($\times 3H_2O$, Na_2 salt)
3′,5′-cAMP	256	14.5	15.0	329.2	347.2 ($\times H_2O$)
5′-CMP	271	9.0	7.6	323.2	323.2
5′-CDP	271	9.2	7.5	403.1	469.1 (Na_3 salt)
5′-CTP	271	9.1	7.5	483.1	549.1 (Na_3 salt)
5′-GMP	252	13.7	11.8	363.2	443.2 ($\times H_2O$, Na_2 salt)
5′-GDP	252	13.7	11.8	443.1	460.9 (Li_3 salt)
5′-GTP	252	13.7	11.7	523.2	541.0 (Li_3 salt)
3′,5′-cGMP	252	13.7		345.2	367.2 (Na salt)
5′-dTMP	267	9.6		322.2	398.2 ($\times 3H_2O$, Na salt)
5′-dTDP	267	9.6		403.4	475.2 ($\times 3H_2O$, Li_3 salt)
5′-dTTP	267	9.6		482.2	624.2 ($\times 3H_2O$, Na_4 salt)
5′-UMP	262	10.0	9.9	324.2	368.2 (Na_2 salt)
5′-UDP	262	10.0	9.9	404.1	470.1 (Na_3 salt)
5′-UTP	262	10.0	9.9	484.1	550.1 (Na_3 salt)
5′-UDP-glucose				544.3	610.3 (Na_3 salt)
NAD	260	18.0	18.0	663.4	705.4 ($\times 2H_2O$, Li salt)
NADH	259	16.9	16.9	745.4	833.4 (Na_4 salt)
	339	6.22			
NADP	260	18.0	18.0	743.4	787.4 (Na_2 salt)
NADPH	259	16.9	16.9	745.4	833.4 (Na_4 salt)
	339	6.2			
FAD	263	38.0	37.0	785.6	807.6 (Na salt)

8.8 Detergents ("Surfactants")

References

Bjerrum OF, Larsen KP, Wilken M (1983) In: Tscheche H (ed.) Modern methods in protein chemistry. Springer, Berlin, p 79
Brito RMM, Vaz WLC (1986) Anal Biochem 152:250
Helenius A, Simons K (1975) Biochim Biophys Acta 415:29
Lichtenberg D, Robson RJ, Dennis EA (1983) Biochim Biophys Acta 737:285
Neugebauer JM (1990) Meth Enzymol 182:239

Table 8.16. Detergents

Name	M_r	cmc[a]	N[b]	Type[c]
Aerosol OT (dioctylsulfosuccinate, sodium salt; sodium bis(2-ethylhexyl)-sulfosuccinate)	444.6	2.5		a
ASB-C6F (4-n-hexylbenzyloylamido-propyldimethylammonium sulfobetain)	411.6			z
ASB-C8F (4-n-octylbenzyloylamido-propyl-dimethylammonium sulfobetain)	439.7			z
Brij 35 (Laureth-23; C12E23)	1200	0.1	40	n
C12E8 (octa(ethylenglycol)-dodecylether; Atlas G2127)	539.1	0.09	120	n
Cetylpyridinium chloride (hexadecylpyridinium chloride; Cetamium)	340.0	0.9	95	c
Cetyltrimethylammonium bromide (CTAB)[d]	364.5	0.9	61	c
CHAPS (3-[(3-cholamidopropyl)-dimethyl-ammonio]-1-propanesulfonate)	614.9	6.5	10	z
Cholate, sodium salt	430.5	16	3	a
1-Decanoyl-phosphatidylcholin (C10-lysolecithin)	447.9	8		z
Deoxy BigCHAP (N,N-bis-(3-D-gluconamidopropyl)-doxycholamide)	862.1	1.4	7	n
Deoxycholate, sodium salt (NaDOC)	414.6	4	10	a
Digitonin	1229.3	0.7	60	n
n-Dodecylglucoside (1-O-n-dodecyl-β-D-glucopyranoside)	348.5	0.13		n
n-Dodecyl-β-D-maltoside	510.6	0.15	98	n
Dodecylsulfate, lithium salt[e]	272.2	8		a
Dodecylsulfate, sodium salt (SDS; sodium laurylsulfate)	288.4	8.2	62	a
Genapol X-080 (11-methyldodecyl-octa(ethylenglycol)ether)	552.8	0.1		n
Hyamin 10X (benzyl-N,N-dimethyl-N-[4-(1,1,3,3-tetramethyl-butyl)-phenoxyethyl]-ammonium bromide)	492.5			c
LPD-12 (Ac-AO(C_{12})AEAAEKAAKYAAEAAEKAAKAO(C_{12})A-NH_2)	23 100	0.2		z
Lysolecithin	$\approx$ 500	$<$ 0.1		a
MEGA-10 (N-(D-gluco-1,3,4,5,6-pentahydroxyhexyl)-hexyl)-N-methyl-deceneamide; decanoyl-N-methylglucamide)	349.5	7		n

Table 8.16. (continued)

Name	M_r	cmc[a]	N[b]	Type[c]
NP 40 (comparable with Triton X-100; nonylphenoxy-polyethoxyethanol; Nonidet P-40)	603.0	0.3		n
n-Octylglucosid (1-O-n-octyl-β-d-glucopyranoside)	292.4	20-25	84	n
1-Palmitoyl-phosphatidylcholin (C16-lysolecithin)	496.6			z
Pluronic F-68 (polyethoxy-polypropoxy copolymere; Synperonic)	8350	0.05		n
Thesit (nona(ethylenglycol)-dodecylether)	582.9	0.1		n
TOPPS (sodium 3-[4-(1,1,3,3-tetramethyl)-phenoxy]-propylsulfonate) [f)]	350.5	4.5		a
Triton X-100 ([4-(1,1,3,3-Tetramethyl-butyl)-phenyl]-deca(ethyleneglycol)ether; tert-C8ΦE9,6; TX 100; comparable with NP40)	628.0	0.25	140	n
Triton X-114 ([4-(1,1,3,3-tetramethyl-butyl)-phenyl]-octa(ethylenglycol)ether; tert-C8ΦE7-8; TX 114)	536.0	0.17		n
Tween 20 (5-eicosa(ethylenglycol)oxy-1-dodecyl sorbitane)	1227.5	0.05		n
Tween 80 (Polysorbate 80)	1309.7	0.01	80	n
n-Undecyl-β-D-maltoside	496.6	0.6		n
Zwittergent 3-12 (Sulfobetain SB 3-12; N,N-dimethyl-N-dodecyl-3-propane sulfonate)	336.6	3.5	55	z
Zwittergent 3-14 (Sulfobetain SB 14; N,N-dimethyl-N-tetradecyl-3-propane sulfonate)	363.6	0.25	83	z

[a] cmc, critical micelle concentration in mMoles/l at 25 °C and maximal 50 mM Na^+. Below cmc, which is depent on temperature and ionic strength, no micelles are formed. Because micelles and monomers are in equilibrium, especially detergents with high aggregation numbers N, e.g. Triton X100, are practically not dialyzable at a concentration above the cmc.

[b] Types of detergents: n, non-ionic; a, anionic; c, cationic; z, zwitterionic

[c] N, Aggregation number (average number of detergent molecules per micelle)

[d] to use in PAGE instead of SDS; direction of migration of protein-detergent complexes + → − (Akin DT, Shapira R, Kinkade JM jr. (1985) Anal Biochem 145:170)

[e] does not precipitate in the cold even at higher concentrations in contrast to SDS

[f] removable with b-cyclodextrin for renaturation of proteins (Rozema D, Gellman SH (1996) J Biol Chem 271:3478)

8.9 Refractive Index and Density of Sucrose Solutions

Table 8.17. Concentration, density ϱ, and refractive index n (20 °C, sodium light source) of sucrose solutions

Sucrose (%) (w/w)	(w/v)	Molarity M	$\varrho_{20°C}$ (g/ml)	$\varrho_{0°C}$ (g/ml)	$n_{20°}^{Na}$
0	0.0	0.00	0.9982	0.9998	1.3330
2	2.0	0.06	1.0021	1.0080	1.3359
4	4.1	0.12	1.0139	1.0162	1.3388
6	6.1	0.18	1.0219	1.0244	1.3418
8	8.2	0.24	1.0299	1.0328	1.3448
10	10.4	0.30	1.0381	1.0413	1.3479
12	12.6	0.37	1.0465	1.0498	1.3510
14	14.8	0.43	1.0549	1.0586	1.3541
16	17.0	0.50	1.0635	1.0674	1.3573
18	18.3	0.56	1.0721	1.0764	1.3606
20	21.6	0.63	1.0810	1.0855	1.3639
22	24.0	0.70	1.0899	1.0947	1.3672
24	26.4	0.77	1.0990	1.1040	1.3706
26	28.8	0.84	1.1082	1.1135	1.3740
28	31.3	0.91	1.1175	1.1231	1.3775
30	33.8	0.99	1.1270	1.1328	1.3811
32	36.4	1.06	1.1366	1.1427	1.3847
34	39.0	1.14	1.1463	1.1527	1.3883
36	41.6	1.22	1.1562	1.1628	1.3920
38	44.3	1.29	1.1663	1.1731	1.3958
40	47.1	1.37	1.1764	1.1835	1.3997
42	49.8	1.46	1.1868	1.1941	1.4036
44	52.7	1.54	1.1972	1.2048	1.4076
46	55.6	1.62	1.2079	1.2156	1.4117
48	58.5	1.71	1.2186	1.2266	1.4158
50	61.5	1.80	1.2296	1.2377	1.4200
52	64.5	1.88	1.2406	1.2490	1.4242
54	67.6	1.97	1.2519	1.2260	1.4285
56	70.7	2.07	1.2632	1.2720	1.4329
58	73.9	2.16	1.2748	1.2837	1.4373
60	77.2	2.26	1.2865	1.2956	1.4418
62	80.5	2.35	1.2983	1.3076	1.4464
64	83.9	2.45	1.3103	1.3198	1.4509
66	87.3	2.55	1.3224	1.3324	1.4558
68	90.8	2.65	1.3347	1.3446	1.4605
70	94.3	2.75	1.3472		1.4651
72	97.9	2.86	1.3598		1.4700
74	101.6	2.97	1.3725		1.4749
76	105.3	3.08	1.3854		1.4799

8.10 Ammonium Sulfate Saturation Table

Ammonium sulfate is one of the cheapest and smoothest agents for precipitation of proteins. It is suitable for protein concentration as well as for protein fractionation.

For description of the amount of ammonium sulfate used for precipitation mostly the term "% saturation" is used, because at a given temperature the concentration of a saturated ammonium sulfate solution is constant. But because the concentration of saturated solution is temperature dependent, the description should also contain data for temperature of saturated ammonium sulfate solution used for precipitation as well as for sample solution.

Tables 8.19 (25 °C) and 8.20 (0 °C) show the amount of solid ammonium sulfate needed to bring up a solution with lower content to a higher percentage of saturation.

Table 8.18. Concentration of ammonium sulfate in saturated solutions at different temperatures

Temperature (°C)	Molarity (moles/l)	Percent		Density (g/cm^3)
		(*w*/*w*)	(*w*/*v*)	
0	3.9	41.42	51.48	1.243
10	3.97	42.22	52.52	1.244
20	4.06	43.09	53.65	1.245
25	4.1	43.47	54.12	1.247
30	4.13	43.85	54.59	1.249

References

Dawson RM, Elliot DC, Elliot WH, Jones KM (1986) Data for biochemical research, 3rd ed. Clarendon Press, Oxford, p 537

Table 8.19. Ammonium sulfate saturation at 25 °C

F =	10	20	30	40	45	50	55	60	65	70	75	80	90	100	%
S (%)	g ammonium sulfate needed per liter														
0	56	114	176	243	277	313	351	390	430	472	516	561	662	767	g
5	27	85	140	210	248	286	318	355	400	441	480	522	618	715	g
10		57	118	183	216	251	288	326	365	406	449	494	592	694	g
20			59	123	155	189	225	262	300	340	382	424	520	619	g
30				62	94	127	162	198	235	273	314	356	449	546	g
40					31	63	97	132	168	205	245	283	375	469	g
45						32	65	99	134	171	210	250	339	431	g
50							33	66	101	137	176	214	302	392	g
60									34	69	105	143	227	314	g
70											35	72	153	237	g
75												36	115	198	g
80													77	157	g

S starting percentage of saturation of ammonium sulfate, F final percentage of saturation

Table 8.20. Ammonium sulfate saturation at 0 °C

F =	10	20	30	40	45	50	55	60	65	70	75	80	90	100	%
S (%)	g ammonium sulfate needed per liter														
0	55	107	166	229	262	295	331	366	404	442	483	523	611	707	g
5	26	80	139	200	232	266	300	336	373	411	450	491	578	671	g
10		54	111	171	203	236	270	305	342	379	418	458	544	636	g
20			56	84	115	145	177	210	244	280	316	392	476	565	g
30				57	87	119	150	184	217	253	289	328	408	495	g
40					29	59	90	122	155	190	225	262	340	424	g
45						29	60	91	125	158	193	229	306	388	g
50							30	61	93	127	161	197	272	353	g
60									31	63	96	131	204	283	g
70											32	66	136	212	g
75												32	102	176	g
80													68	141	g
90														71	g

S starting percentage of saturation of ammonium sulfate, F final percentage of saturation

8.11 Diluted Solutions

Table 8.21 presents the volume of starting solution A, which gives the wanted concentration after filling up to 1000 ml.

Table 8.21. Preparation of diluted solutions

Substance	Conc.	Soln. A (ml) per 1000 ml	Density ϱ (20 °C)
Acetic acid	A = 99% (w/w)		1.052
	1 M	57.8	1.007
Ammonia	A = 25% (w/w)		0.907
	10% (w/w)	422.3	0.958
	5% (w/v)	215.4	0.977
	1 M	75.1	0.992
Formic acid	A = 90% (w/w)		1.205
	1 M	42.4	1.010
Hydrochlorid acid	A = 36% (w/w)		1.180
	25% (w/w)	694.4	1.124
	10% (w/w)	277.7	1.047
	1 M	85.9	1.015
Nitric acid	A = 65% (w/w)		1.391
	10% (w/w)	116.6	1.054
	1 M	69.6	1.032
Perchloric acid	A = 70% (w/w)		1.674
	10% (w/w)	90.5	1.060
	1 M	85.7	1.058
Phosphoric acid	A = 90% (w/w)		1.746
	10% (w/w)	90.4	1.060
	1 M	56.1	1.051
Sulfuric acid	A = 98% (w/w)		1.836
	25% (w/w)	167.1	1.178
	10% (w/w)	63.0	1.066
	5% (w/w)	29.3	1.032
	1 M	54.5	1.061

8.12 Mixture Rule

If a diluted solution shall be made from a higher concentrated solution and a lower concentrated one, or if you want to know which concentration a substance has after mixture of two solutions, the mixture rule is a simple and fast method to calculate the required data.

If the difference of densities of the solution is relatively high, you obtain better results if using densities of participating solutions instead of concentrations.

The mixture rule expressed as an equation is:

$$V_1 = F - S_2 \quad \text{and} \quad V_2 = S_1 - F$$

S_1: starting concentration of the more concentrated solution; S_2: starting concentration of the less concentrated solution; F: wanted final concentration; V_1: needed volume of the more concentrated solution; V_2: needed volume of the less concentrated solution

Example

How many volumes of a 37% hydrochloric acid (starting concentration S_1 HCl = [HCl] = 37%) have to be mixed with volumes of water (starting concentration S_2 [HCl] = 0%) to get a 10% hydrochloric acid? (It is more precise to take the densities instead of concentrations, especially when there is a high difference in specific gravity as shown in this example.)

Result

37		10		1.180		0.047
↘		↗		↘		↗
	10		and		1.047	
↗		↘		↗		↘
0		27		1.000		0.133
	concentration				density	

Ten volumes of 37% HCl (V_1) have to be mixed with 27 vol. of water (V_2) (from density: 28.3 vol.).

9 Statistics and Data Analysis

9.1 Statistical Equations

Some equations frequently used in evaluation of experimental data are given in this chapter. Their theoretical background and mathematical derivations can be found in cited literature.

9.1.1 Mean and Related Functions

The arithmetic mean m of a population of values x with n members is given by:

$$m = \frac{\sum_{i=1}^{n} x_i}{n}$$

Degree of freedom of mean F is:

$$F = n - 1$$

Standard deviation (s or SD) of mean[1] quantifies variability:

$$s = \sqrt{\frac{\sum_{i=1}^{n} (x_i - m)^2}{n-1}} = \sqrt{\frac{\sum_{i=1}^{n} (x_i - m)^2}{F}} = \sqrt{\frac{\sum_{i=1}^{n} x^2 - \left(\sum_{i=1}^{n} x\right)^2 / n}{n-1}}$$

Standard error of mean (SEM):

$$SEM = \frac{s}{\sqrt{n}}$$

The standard deviation quantifies how much the values vary from one another, whereas SEM quantifies how precisely you know the true mean. If you want to show the variability of biological values, use SD. For characterizing of the precision of an in vitro experiment, SEM is the best expression.

The coefficient of variation cv (relative variability) is calculated by

$$cv = \frac{SD}{m}$$

[1] has the same dimension as the respective mean

The absolute error of mean[1] EM is calculated from formula

$$EM = \left| \frac{t \cdot SD}{\sqrt{n}} \right|$$

(for calculation of t see Sect. 9.1.3).

If the observed data follow a bell-shaped GAUSSIAN (normal) distribution, the confidence interval CI of mean is defined by

$$CI_{1-P} = m_x \pm EM$$

N: number of data x_i; P: probability of significant deviation of a value from data within interval confidence; t: – borders of the $(1-P)$th part of the area of the GAUSSIAN distribution for a given F.

Mostly the 0.95 confidence interval of mean is given, i.e., you can be 95% sure that your randomly selected sample of a population is included in the population mean.

To calculate EM and CI at a given F with distinct probability P take t from Table 9.1.

9.1.2 Correlation: Linear Regression

Assuming that the variable y depends on the variable x by a linear function

$$y = a + b \cdot x\,,$$

it is often observed that the measured y values randomly deviate from (theoretical) values calculated from the equation. Because the true y values are unknown, it is possible to fit the observed values by the method of least squares. Calculation of the regression coefficients a and b by this method gives a regression line. The correlation coefficient r serves as a measure for the goodness of fit[2].

$$b = \frac{\sum_{i=1}^{n} (x_i - m_x) \cdot (y_i - m_y)}{\sum_{i=1}^{n} (x_i - m_x)^2}$$

$$a = \frac{\sum_{i=1}^{n} y_i - b \cdot \sum_{i=1}^{n} x}{n} = m_y - b \cdot m_x$$

$$r = \frac{1}{n-1} \cdot \frac{\sum_{i=1}^{n} (x_i - m_x) \cdot (y_i - m_y)}{s_x \cdot s_y}$$

[2] $1 \geq |r| \geq 0.8$, very well correlation; $0.8 \geq |r| \geq 0.6$, medium correlation; $|r| < 0.6$, bad or no correlation

Table 9.1. The *t* values for *P* = 0.1, 0.05, 0.02, 0.01, and 0.001

F	P = 0.10	0.05	0.02	0.01	0.001
1	6.31	12.71	31.82	63.66	318.31
2	2.92	4.30	6.97	9.92	22.23
3	2.35	3.18	4.54	5.84	10.21
4	2.13	2.78	3.75	4.03	7.17
5	2.01	2.57	3.37	4.03	5.89
6	1.94	2.45	3.14	3.71	5.21
7	1.89	2.36	3.00	3.50	4.79
8	1.86	2.31	2.90	3.36	4.50
9	1.83	2.26	2.82	3.25	4.30
10	1.81	2.23	2.76	3.17	4.14
11	1.80	2.20	2.72	3.11	4.03
12	1.78	2.18	2.68	3.05	3.93
13	1.77	2.16	2.65	3.01	3.85
14	1.76	2.14	2.62	2.98	3.79
15	1.75	2.13	2.60	2.95	3.73
16	1.75	2.12	2.58	2.92	3.69
17	1.74	2.11	2.57	2.90	3.65
18	1.73	2.10	2.55	2.88	3.61
19	1.73	2.09	2.54	2.86	3.58
20	1.72	2.09	2.53	2.85	3.55
21	1.72	2.08	2.52	2.83	3.53
22	1.72	2.07	2.51	2.82	3.51
23	1.71	2.07	2.50	2.81	3.49
24	1.71	2.06	2.49	2.80	3.47
25	1.71	2.06	2.49	2.79	3.45
26	1.71	2.06	2.48	2.78	3.44
27	1.70	2.05	2.47	2.77	3.42
28	1.70	2.05	2.47	2.76	3.41
29	1.70	2.05	2.46	2.76	3.40
30	1.70	2.04	2.46	2.75	3.39
40	1.68	2.02	2.42	2.70	3.31
60	1.67	2.00	2.39	2.66	3.23
120	1.66	1.98	2.36	2.62	3.16

x_i and y_i: values of the ith pair of data; m_x and m_y: mean of x and y values, respectively; n: number of data pairs; s_x and s_y: standard deviation of the x and y values, respectively.

If there is a linear connection, calculation of unknowns by the regression function is possible. But as shown in Fig. 1.1, for example, the linear correlation very often exists only approximately in a small range.

Computation of a linear regression is possible with each population of x–y pairs of data. But it should be checked whether other functions may be the basis of the connection between x and y, es-

Comparison of different fits

pecially when a low correlation coefficient r is obtained. If there is no linear correlation, a non-linear regression calculation of a hyperbola or sigmoid function should be selected. To support the decision for another hypothetical model, some computer programs offer a comparison of the different fits.

9.1.3 The *t*-test (Student's Test)

Null hypothesis

During an experiment one has a set of data which seems to be formed from two groups. Is it wise to calculate a single mean of all data because difference between means of subpopulations results from simple fluctuations, or have we to form two subpopulations with significantly different means? The hypothesis that all data are parts of a sole population is the so-called null hypothesis. The rejection of the null hypothesis leads to establishing a significant difference between both groups of data. The *t*-test is used to decide this question.

Student's test compares the means of both groups and is restricted to two sets of data, which should come from a normal distribution. The *t*-test works well if the variances of both groups are not very different and the number of data in each data set is similar.

If the null hypothesis is true, the calculated *t*-value is smaller or equal to that obtained from Table 9.1 for the respective degree of freedom and a chosen probability P, i.e., there is a significant difference between the both means if $t_{calculated} > t_{1-P(Table)}$.

For example, if the calculated *t*-value of group A ($n = 9$) and B ($n = 10$) with $F = F_A + F_B = 17$, is 2.70, the groups are significantly different at $P = 0.02$, but they are not different at $P = 0.01$.

The probability for a significant difference is $(1-P) \cdot 100 = 98\%$.

$$t = (m_A - m_B) \cdot \frac{\sqrt{\frac{n_A \cdot n_B}{n_A + n_b}}}{s_D}$$

or, if $n_A = n_B$

$$t = \frac{(m_A - m_B)}{s_D} \cdot \sqrt{\frac{n}{2}}$$

The standard deviation s_D of the difference of the both means is calculated by the equation

$$s_D = \sqrt{\frac{\sum (x_{Ai} - m_A)^2 + \sum (x_{Bi} - m_B)}{F}}$$

with

$$F = n_A + n_B - 2$$

and

$$m_A = \frac{\sum_{n=1}^{i} x_{Ai}}{n_A}$$

in addition to

$$m_B = \frac{\sum_{n=1}^{i} x_{Bi}}{n_B}$$

x_{Ai} and x_{Bi}: data of group A and B, respectively; m_A and m_B: mean of the respective values; n_A and n_B: number of data of the respective groups; F: degree of freedom

Prerequisite for the *t*-test is a normal distribution of data, i.e., the frequencies of data with the same deviation from mean forms a bell-shaped curve. In case of a large number of experimentally obtained data, mostly a GAUSSIAN distribution is given.

In practice, P is given as a threshold for statistical significance and a low P is read as more significant than a higher value. In a strong sense, this definition is not correct, but this interpretation is common usage. Table 9.2 gives P values and their explanation as found very often in scientific literature.

References

Armitage P, Berry G, Matthews JNS (2002) Statistical methods in medical research, 4th ed. Blackwell, Malden Massachusetts

Dawson BD, Trapp RG (2000) Basic and clinical biostatistics. Appleton and Lange

Motulsky H (1995) Intuitive biostatistics. Oxford University Press, Oxford

Wardlaw AC (1987) Practical statistics for experimental biologists. Wiley, Chicheser

Table 9.2. Steps of significance

P value	Verbal expression	Symbol
< 0.001	Extremely significant	***
0.001 – 0.01	Very significant	**
0.01 – 0.05	Significant	*
> 0.05	Not significant	n.s.

9.2 Data Analysis

9.2.1 Receptor–Ligand Binding

Computation of data obtained by enzyme kinetic experiments, or receptor binding studies using sophisticated software, are state of the art. But understanding the transformations of these data into linear correlations, as well as plots of these transformations, is necessary for critical reading of papers, on the one hand, and for testing results of computer programs, on the other.

The main plots used in enzyme kinetics and receptor binding studies are the Scatchard plot, the Lineweaver–Burk plot, and the linearization for estimation of the Hill coefficient. This chapter gives a short survey of these transformations of enzyme kinetics or receptor binding data.

The interaction of reversibly binding ligand L (enzyme substrate) with its receptor R (enzyme) follows the law of mass action:

$$K_D = \frac{k_1}{k_2} = \frac{[R] \cdot [L]}{[RL]}$$

with K_D, dissociation constant; [R], concentration of receptor; [L], concentration of ligand; and [RL], concentration of the receptor–ligand complex.

If a distinct amount of receptor is incubated with its ligand, a part of the ligand will be bound to the receptor in a proportion given by the equilibrium ratio. The concentration of the bound portion B of the total ligand concentration L is equal to the concentration of the receptor–ligand complex:

$$[RL] = [B] = [L] - [F]$$

[F] gives the concentration of free, unbound ligand.

Merging both equations and transformation of the result gives the Scatchard graph, characterized by plotting [B]/[F] on ordinate and $1/K_D$ on abscissa. The constant B_{max} represents that concentration of L needed for complete saturation of all binding sites at the receptor and the maximal number of binding sites, respectively.

$$\frac{[B]}{[F]} = \frac{1}{K_D} \cdot \left([B] + B_{max}\right)$$

An example for binding experiment is given in Protocol 5.3.2.2.

The Hill coefficient n gives an impression of the number of binding sites for a ligand per single receptor molecule, i.e., if $n = 1$, the receptor has one binding site for the specified ligand. Using a further transformation, the slope of the straight line is the Hill coefficient:

$$\lg\left(\frac{[B]}{B_{max}}\right) = n \cdot \lg[F] - \lg K_D$$

The association constant k_1 is calculated either by use of the independent determined maximal number of binding sites B_{max} (v_{max} in enzyme kinetics) from the equation

$$k_1 = \frac{1}{t \cdot (L_T - B_{max})} \cdot \ln \frac{B_{max} \cdot (L_T - B)}{L_T \cdot (B_{max} - B)}$$

or, if the dissociation constant k_2 is known, from the plot of the functions

$$\ln \frac{B_e}{B_e - B} = k_{obs} \cdot t$$

and

$$k_{obs} - k_2 = k_1 L_T$$

B_e: concentration of the bound ligand at equilibrium; B_0: concentration of the specifically bound ligand at time t = 0; B: concentration of the specifically bound ligand at time t; L_T: total ligand concentration; k_{obs}: experimentally determined rate constant; t: time

The dissociation constant k_2 is determined by off-kinetics experiments. It is the slope of the function

$$\ln \frac{B}{B_0} = k_2 \cdot t$$

The dose-dependent saturation of a constant amount of receptor is described by sigmoid or hyperbolic curve shape. If the measuring signal is plotted against ligand concentration, the curves show a minimal signal (blank, lower plateau) and an upper plateau (B_{max}). The equation for the hyperbola is

$$y = [\text{bound}]\text{-}[\text{blank}] = \frac{B_{max} \cdot x}{K_D + x} = \frac{B_{max} \cdot [L_{total}]}{K_D + [L_{total}]}$$

or in the case of two binding sites with different binding characteristics (different dissociation constants K_D and MICHAELIS–MENTEN constant K_M, respectively)

$$y = \frac{B_{max1} \cdot x}{K_{D1} + x} + \frac{B_{max2} \cdot x}{K_{D2} + x}$$

In these equations the abscissa value y represents the concentration of bound ligand, whereas x is the total concentration of ligand.

A sigmoid shape is described by the equation

$$y = \text{blank} + \frac{B_{max} - \text{blank}}{1 + 10^{(\lg(EC_{50}) - x) \cdot n}}$$

EC_{50}: concentration halfway between blank and B_{max}; n: Hill coefficient; x: logarithm of ligand concentration and substrate concentration; y: amount of bound ligand: ("bound"-"blank")

If two binding sites with different affinities are present, visible in a double-sigmoid curve, the equation is modified as follows:

$$y = \text{blank} + \frac{B_{max_h} \cdot 10^{n_h \cdot x}}{10^{n_h \cdot lgEC_{50_h}} + 10^{n_h \cdot x}} + \frac{B_{max_l} \cdot 10^{n_l \cdot x}}{10^{n_l \cdot lgEC_{50_l}} + 10^{n_l \cdot x}}$$

(h and l indicate the respective values for high-affinity and low-affinity binding sites[3]).

Several computer programs have the opportunity to compare the goodness of fit obtained by non-linear regression calculations of different models. This algorithm uses a so-called F test to quantify the sum of squares of both fits and allows to know which function is more appropriate for your data. So it is possible to decide whether only one binding site or two equal sites or two sites with different affinities are occupied by the ligand. The function types for hyperboloid or sigmoid binding characteristics given above must not be compared by the F test because in the first case concentrations are used and in the latter logarithms of concentrations are taken.

References

Wells JW (1992) In: Hulme EC (ed.) Receptor–ligand interactions: a practical approach. IRL Press, Oxford, p 289

Motulsky HJ (1999) GraphPad Prism, version 3.0. GraphPad Software, San Diego, p 303

9.2.2 Enzyme Kinetics

It is impossible to describe and explain enzyme kinetics unless is explained by an entire book; therefore, this chapter describes only briefly some aspects. It is strongly recommended to read once more a textbook on enzymology and enzyme kinetics. Especially the reaction kinetics of enzyme oligomeres, multi-enzyme complexes, and phenomena of cooperation are too complex to explain in just a few pages.

If enzymes are described under the aspect of reaction mechanisms, the maximal rate of turnover v_{max}, the MICHAELIS and MENTEN constant K_M, the half maximal inhibitory concentration IC_{50}, and the specific enzyme activity are keys of characterization of the biocatalyst. Even though enzymes are not catalysts in a strong chemical sense, because they often undergo an alteration of structure or chemical composition during a reaction cycle, theory of enzyme kinetics follows the theory of chemical catalysis.

In the most simple case an enzymatic reaction is described by the equation

$$E + S \rightleftarrows ES \rightleftarrows EP \rightleftarrows E + P$$

[3] According to ZERNIG et al. (1994) J Pharmacol Expt Therap 269:57

These equilibriums are subjected to the law of mass action:

$$K_a = \frac{k_1}{k_{-1}} = \frac{[ES]}{[E] \cdot [S]}$$

K_a is the association equilibrium constant and is inverse proportional to the dissociation equilibrium constant K_D:

$$K_D = \frac{1}{K_a} = \frac{k_{-1}}{k_1} = \frac{[E] \cdot [S]}{[ES]}$$

[E]: enzyme concentration; [S]: substrate concentration; [ES]: concentration of the enzyme-substrate complex; [P]: product concentration; k_1: association equilibrium rate constant; k_{-1}: dissociation equilibrium rate constant (rate constant of the back reaction).

It is stated that during an in vitro enzymatic reaction the concentration of the enzyme shall not change during the test, and that the substrate concentration exceeds the enzyme concentration in orders of magnitude: in a first approximation the substrate concentration is practically constant, too. Both of these assumptions transform a reaction of 2nd order into the much simpler reaction of 0th order. If the concentrations of enzyme and substrate are similar, we get a reaction of 1st order. The reaction rate v for the association reaction

$$E + S \rightarrow ES$$

is in the case of large substrate excess (0th order)

$$v = \frac{d[S]}{dt} = k_1$$

and

$$v = -\frac{d[S]}{dt} = \frac{d[P]}{dt} = k_1 \cdot [S]$$

respectively. Integration of the differential quotients over time $t = 0$ to t gives

$$\ln[S] = \ln[S_0] - k_1 \cdot t$$

$[S_0]$: substrate concentration at $t = 0$

This equation allows the determination of the rate constant of the association reaction, and analogously, by measuring the product forming, the dissociation rate constant.

A further important value is the time of half-change $_{t1/2}$, i.e., that time at which half the amount of substrate is converted ($[S] = 0.5 \cdot [S_0]$):

$$t_{1/2} = \frac{\ln 2}{k} = \frac{0.69315}{k}$$

In steady state, i.e., when association and dissociation occur at the same speed, the change of [ES] and [E] are the same in the time interval. Because $[E_0]$, the enzyme concentration at $t = 0$, is the sum of [E] plus [ES], and because the substrate conversion is maximal when enzyme molecules are saturated with substrate and involved within the catalytic process ($v_{max} = k_3 \cdot [E_0]$), it is possible to transform the rate equation into the MICHAELIS–MENTEN equation

$$v = \frac{v_{max} \cdot [S]}{K_M + [S]}$$

K_M is the MICHAELIS constant or the enzymatic reaction and is defined by

$$K_M = \frac{k_2 + k_3}{k_1}$$

If the reaction rate v is plotted against the substrate concentration [S] (measuring v at different substrate concentrations within a linear range of substrate conversion), v_{max} is calculated for $[S] \rightarrow \infty$ and the value of $0.5 \cdot v_{max}$ gives K_M. If $[S] \gg K_M$, then v becomes v_{max}.

K_M and v_{max} are mostly determined from linearized plots derived from conversions of the MICHAELIS–MENTEN equation. Some of these linearizations are given in Table 9.3.

The LINEWEAVER–BURK plot is very useful for descriptions of type and effects of inhibitors: Competitive inhibitors have the same intercept on the ordinate and different intercepts on abscissa, non-competitive inhibitors give the same intercept at the abscissa but different at the ordinate. In the case of (partially) inhibited reactions, the slope is larger than at the respective non-inhibited reaction.

If the effect of an inhibitor on an enzyme is to be investigated, the DIXON plot is recommended. To obtain data for the DIXON plot, estimate the reaction rate at constant substrate concentration and vary the inhibitor concentration [I]. At competitive inhibition, all the obtained straight lines coincide at a point with the coordinates $x = -K_I$, $y = 1/v_{max}$, and at non-competitive inhibition all the straight lines have the same intercept on abscissa at $x = -K_I$. At

Table 9.3. Transformations for graphic determination of K_M and v_{max}

Plot according to	Plot on		Intercept with the		Slope
	Abscissa	Ordinate	Abscissa	Ordinate	
LINEWEAVER and BURK	1/[S]	1/v	$-1/K_M$	$1/v_{max}$	K_M/v_{max}
HANES	[S]	[S]/v	$-K_M$	K_M/v_{max}	$1/v_{max}$
EADIE and HOFSTEE	v/[S]	v	v_{max}/K_M	v_{max}	$-K_M$
DIXON	[I]	1/v		$1/v_{max}^{a}$	

[a] At high excess of substrate

large substrate excess $1/v_{max}$ is obtained from the ordinate intercept of the straight line.

The specific enzyme activity is defined as the amount of converted substrate and formed product, respectively, per time unit and amount of enzyme at defined pH, temperature, and buffer composition. The specific activity is given as arbitrary units (e.g., units/mg/min of units/O.D./min; the international unit IU is defined as the conversion of 1 µmol substrate and forming of 1 µmol product, respectively, per minute) or as SI unit "kat/mg" (Mol/s/kg).

The value "IC_{50}" is that concentration of inhibitor, which reduces the enzyme activity to 50% of the activity in the absence of inhibitors.

References

Bisswanger H (1994) Enzymkinetik. Theorie und Methoden. 2. neubearb. Aufl., VCH, Weinheim
Eisenthal R, Danson MJ (eds.) (1992) Enzyme assays: a practical approach. IRL Press, Oxford

9.2.3 Determination of Molecular Mass by SDS-PAGE

The independent determination of molecular masses by SDS-PAGE is impossible. To estimate the molecular mass of a protein, measure the path of that protein or calculate its R_f value (distance of the protein from origin/distance of electrophoresis front from origin) and compare these values with that of marker proteins, i.e., proteins with independently determined molecular masses. This method is successful only if the protein of interest behaves regularly in SDS-PAGE, i.e., it is totally unfolded by SDS, has a rod-like shape, and the SDS/protein ratio is the same for the unknown and the marker protein. Especially highly hydrophobic proteins and glycoproteins often deviate from these assumptions.

Calculation of the molecular mass of an unknown protein follows the same procedure as, for example, quantitative protein determination: plotting of the R_f of the calibration proteins against their molecular mass; computing of a standard curve and estimation of the MW of the unknown protein; and using the regression functions of the standard curve (c.f. Fig. 2.1).

To be sure that the obtained molecular mass is correct, a Ferguson plot is recommended: Plot $\log_{10}R_f$ from runs of the analyzed protein in gels with different %T against "%T." Regular proteins give a straight line; glycoproteins often give a hyperbolic curve resulting in too low molecular mass obtained from runs in gels with higher %T and/or %C.

References

Hames BD(1990) In: Hames BD, Rickwood D (eds.) Gel electrophoresis of proteins: a practical approach, 2nd ed. IRL Press, Oxford, p 16

9.3 Diagnostic Sensitivity and Specificity

The goal of a diagnostic method is to detect all people within a population bearing the disease marker (no false-negative people) and to have no false-positive results, i.e., positive signal from a healthy person. The terms that characterize these demands are

$$\text{Sensitivity} = A/(A + C)$$

and

$$\text{Specificity} = D/(B + D)$$

with

$$\sum \text{“ill”} = A + C$$

and

$$\sum \text{“healthy”} = B + D$$

Further measures are:

$$\text{Positive predictive value:} \qquad \text{PPV} = A/(A + B)$$

$$\text{PPV} = \frac{\text{Sensitivity} \cdot \text{Prevalence}}{\text{Sensibility} \cdot \text{Prevalence} + (1 - \text{Specificity}) \cdot (1 - \text{Prevalence})}$$

and

$$\text{Negative predictive value:} \qquad \text{NPV} = \frac{D}{C + D}$$

$$\text{NPV} = \frac{\text{Specificity} \cdot (1 - \text{Prevalence})}{\text{Specificity} \cdot (1 - \text{Prevalence}) + (1 - \text{Sensitivity}) \cdot \text{Prevalence}}$$

A: right positive; B: false positive; C: false negative; D: right negative.

Diagnostic specificity and sensitivity are analyzed by receiver-operator curves (ROC) using data obtained from defined healthy populations, patients with diseases other than the investigated ones, and patients with clinical relevance to the respective disease.

9.4 Software for the Lab

The following software acts for a lot of offers. The programs listed below are, of course, excellent, but it should be checked if other software is more convenient with respect to distinct advantages or parameters.

Since the development of software is extremely quick, no versions are indicated. Most of the software is available for PC as well as for Macintosh computers.

9.4.1 Data Analysis and Presentation

Prism: http://www.graphpad.com/
GraphPad Software, Inc., 11452 El Camino Real, #215, San Diego, CA 92130, USA

Origin for Windows: http://www.originlab.com/
OriginLab Corporation, One Roundhouse Plaza, Northampton, MA 01060, USA

SigmaPlot: http://www.systat.com/products/SigmaPlot/
Systat Software, Inc., 501 Canal Blvd, Suite E, Point Richmond, CA 94804–2028, USA

9.4.2 Software for Statistics

InStat: http://www.graphpad.com
GraphPad Software, Inc., 11452 El Camino Real, #215, San Diego, CA 92130, USA

MedCalc for Windows - statistics for biomedical research:
http://www.medcalc.be
MedCalc Software, Broekstraat 52, 9030 Mariakerke, Belgium

Statgraphics: http://www.statgraphics.com
StatPoint, Inc., 2325 Dulles Corner Boulevard, Suite 500, Herndon, Virginia 20171, USA

WinStat: http://www.winstat.de
R. Fitch Software, St.-Martin-Allee 1, 79219 Staufen, Germany

9.4.3 Other Software

Reference Manager: http://www.refman.com
Thomson ResearchSoft, 425 Market St., 6th Floor, San Francisco, CA 94105, USA
Personal reference bibliographic management system.

Molecular Drawing and Visualization

ChemWindows: http://www.bio-rad.com
Bio-Rad Laboratories, Informatics Division, 3316 Spring Garden Street, Philadelphia, PA 19104-2596, USA
Drawing of chemical structures and computation of 3D-models of small organic molecules

Chem3D: CambridgeSoft, 100 CambridgePark, Cambridge, MA 02140, USA, www.CambridgeSoft.com
Drawing and computation of organic chemical molecules

RasWin (freeware):
http://www.bernstein-plus-sons.com/software/rasmol
Molecular graphics visualization tool, using Protein Data Bank *.pdb files

RasTop (freeware):
http://www.geneinfinity.org/rastop/download.htm
Molecular graphics visualization tool, using Protein Data Bank *.pdb files

SwissProt PDBViewer (freeware):
http://us.expasy.org/spdbv/mainpage.htm
Swiss-PdbViewer allows to analyze several proteins at the same time.

9.4.4 Selected Internet Links

Publication Databases:

PubMed (search of publications in life sciences): PubMed, a service of the National Library of Medicine:
http://www.ncbi.nlm.nih.gov/entrez/query.fcgi

Current Contents: http://isi15.isiknowledge.com/portal.cgi/?DestApp=WOS&Func=Frame

Medline plus: http://medlineplus.gov

International Patents:

European Patent Office http://ep.espacenet.com
Intellectual Property Digital Library
http://www.wipo.int/ipdl/en
United States Patent and Trademark Office
http://www.uspto.gov/patft

Molecular Databases:

Bioinformatic Harvester: http://harvester.embl.de/GenBank

Brookhaven Protein Database:
(PDB) http://www.pdb.mdc-berlin.de/pdb

Entrez, the life sciences search engine:
http://www.ncbi.nlm.nih.gov/gquery/gquery.fcgi

Swiss-Prot/TrEMBL: http://us.expasy.org/cgi-bin/sprot-search-ful?makeWild=&SEARCH=

Enzymes, Receptors, Ligands, and Substrates:

Enzymes – systematics and properties:
http://www.brenda.uni-koeln.de/index.php4?page=information/all_enzymes.php4?ecno=

Human Protein Reference Database: http://www.hprd.org/protein

PubChem - Chemical structures of small organic molecules and information on their biological activities:
http://pubchem.ncbi.nlm.nih.gov

Search within the archive of websites: http://www.archive.org

Subject Index

1-cyan-4-dimetylaminopyridinium-tetrafluoroborate *see* CDAP
1-napthyl red 50
2D-PAGE 41
3,3′,5,5′-tetramethylbenzidine *see* TMB
3,3′-diaminobenzidine 72
4-(dimethylamino)azobenzene *see* DABITC
4-chloro-1-naphthol 72
6-aminohexanoic acid *see* EAC
8-anilino-1-naphthalenesulfonic acid *see* ANS

absorption 12
  disulfide bond 12
  nucleic acid 12, 16
  peptide bond 12
ABTS 158, 159
activity coefficient 192
additives 126
adjuvant, Freund's 143
agar 151
agarose 45, 151
  low endoosmosis 45
alkaline phosphatase 71, 73, 122, 138, 158
Amido Black 10 B 8, 54
amino acids 223
ampholyte 42
AMPPD 74
ANS 62
antiserum 117
APMSF 221
aqua regia 141
arithmetic mean 233
ascorbic acid 37
association constant 238
avidin 71, 121, 143

β-galactosidase 71, 134, 158
barbital buffer 151, 154
BCIP 73
Beer–Lambert law 11, 12, 21
benzamidine 221
biocide 106
biotin 71, 99, 121, 130, 215
  N-hydroxysuccinimide ester 122
Bismarck Brown 55
blocking 71
  ELISA 158–160
  membrane 71
  reagent 71
BME Hanks 205
BME modified 204
Bolton–Hunter reagent 183, 187, 188
borax 206
Bray solution 189
Brij 35 226
bromophenol blue 27, 29, 33, 36, 38–41, 44, 46, 47, 49, 155
Brönsted definition 198
BSA, activated 136
buffer capacity 195, 196
buffer, volatile 126
buoyant density 177

C.I. *see* Color Index
C12E8 226
calibration buffer 207
calpain inhibitor 221
carbodiimide 115, 130
Carbowax 64, 142
carrier 130, 134
CDAP 113
Centricon 66
Cerenkov radiation 182, 189
cetyltrimethylammonium bromide *see* CTAB
chaotropic substance 93, 108, 111
CHAPS 42, 226
checking number 222

chemiluminescence 74
exposure 75
chloramine T 187
Cibacron Blue 121
Color Index 6
Concanavalin A 75, 76, 114, 117, 136, 137
confidence interval 234
conjugation, reagents 130
conversion factors 211
Coomassie Brilliant Blue 6, 7, 30, 31, 35, 36, 54, 55, 65
copper(II) phthalocyanine-3,4′,4″,4‴-tetrasulfonic acid *see* CPTS
correlation coefficient 234, 236
CPTS 62, 64, 65
cresol red 37
cross-linker 131
crossing over electrophoresis 155
CSPD 74
CTAB 23, 30

DABITC 83
dabsyl chloride 52
DAPI 15
Darcy's law 95
Debye-Hückel relationship 196
degree of activation 114, 119
degree of dissociation 196
degree of freedom 233
degree of substitution 139
densitometry 55
deoxycholate, sodium 124
deoxyribose 19
desorption, biospecific 102, 110
detection, amperometric 106
detergent 5, 226, 227
dextran blue 99
dialysis membrane 66
diethylpyrocarbonate 175
digitonine 5, 226
digoxigenin 76, 130
diluted solutions 231
diphenylamine 14
disc electrophoresis 26
dissociation constant 175, 193, 238, 239
Dixon plot 242
DMEM *see* MEM Dulbecco
dodecylsulfate 226
dye, fluorescent 140

EAC 69
elution buffer 112
epitope 30, 70, 156
influence of binding material 156
equilibrium constant 191
ethidium bromide 15, 46, 140
extrapolation 20

Farmer's reducer 61
Fast Green 55, 63, 64
Ferguson plot 24, 243
Ficoll 166, 168
film remover 45
fish gelatin 71, 159
FITC 139, 140
flow rate 95
linear 95
volumetric 95
fluorescein isothiocyanate *see* FITC
fluorescence quenching 85, 87
Folin–Ciocalteu's phenol reagent 3
food analysis 10
formamide 47
fuchsin, basic 39, 49, 62

Gaussian distribution 234, 237
gel
fixation 31
gradient 24, 28
homogenous 24
polymerization 27, 37
separation 33, 35, 39
separation distance 25
slab 24, 27
stacking 35, 39
gel overlay 45
GelBond 42, 44
genetic code 224
glycine-HCl buffer 201
glycolipid 62
glycoprotein 23, 40, 62, 121
glyoxal 47
goodness of fit 234
GPC
flow rate 98, 99
sample volume 96
gradient
density 178
sucrose 176
gradient elution 105
gradient, density 170

H_2O_2-urea adduct 158, 159

HANES' reagent 85, 87
hapten 130, 143
HENDERSON–HASSELBALCH equation 192, 198
hexose 19
HILL coefficient 238
His_6 tag 123
histones 37
Hoechst 33 258 15
HOFMEISTER series 93, 124
horseradish peroxidase 71, 72, 117, 122, 135, 158
HPIEC 90
hydrazide 76
  digoxigenin 77
hydrophobicity 107
hydrostatic pressure 94, 97

IEC
  capacity 102, 104
  monosaccharides 106
  oligosaccharides 106
  pre-cycling 103
IECsample volume 105
IEF 41
Immobiline 42
immunization scheme 144
immunoelectrophoresis 40
immunoglobulins, absorption coefficient 145
inclusion bodies 91
injection 143
  intradermal 143
  intramuscular 143
  intravenous 143
intensifying screen 80
Iodo-gen 187
iodoacetamide 26, 30, 44, 150, 221
iodoacetate 47
ionic strength 25, 155, 195
IPG-Dalt 41, 42
isoelectric focusing *see* IEF

KJELDAHL factors 10
KLH 130, 132–134, 143
Krebs-Henseleit-Ringer Buffer 206

law of mass action 238
lectin 40, 75
  specifity 75
Leupeptin 221
ligand 109, 237, 238
ligand-ligate interaction 38, 40
ligate 109
linear correlation 20
linear function 234
LINEWEAVER–BURK plot 238, 242
lipid, remoning of 13

marker enzymes 171
mass spectrometry 24
MCA-Gly peptide 132
medium 179
  IMDM 179
  RPMI 179
MEGA-10 226
MEM Dulbecco 205
membrane
  blocking 77
method of least squares 234
methyl red 50
Metrizamide 166, 168
MICHAELIS–MENTEN constant 239, 240
MICHAELIS–MENTEN equation 242
mixture rule 232
modification, posttranslational 83, 91
molality 209
molarity 209
molybdatophosphate complex 17
monosaccharide 19

N-hydroxysuccinimide *see* NHS, 130
N-methylmaleiimide 26
native electrophoresis 40
NBT 73
negative predictive value 244
NEM 30
NHS 117, 122
ninhydrin 48
nitrendipine 173, 174
nitrocellulose 156
normality 209
Northern blot 68, 78
NP 40 227
NTA resin 123
nucleotides 225
null hypothesis 236
Nycodenz 166, 168

o-Phenanthroline 221
o-phenylenediamin *see* OPD
O'FARRELL technique 41

off-kinetics 174, 239
OliGreen 15
on-kinetics 174
OPD 158, 159
Orange G 49
orcin 13, 14
ouabain 170–172
ovalbumin 130, 134, 136

p-nitrophenyl phosphatase 170
p-nitrophenyl phosphate 138, 160
PAGE, continuous 32
papain 149
parabene 106
PBS 203
Pefabloc 221
PEG 126, 128, 174
PEI Cellulose 85
penicillin 179
pentose 19
pepsin 149
Pepstatin A 221
percoll 166, 168, 178
phase 89
  mobile 89, 95
  stationary 89
phenol 19
phenol red 31
phenylisothiocyanate *see* PITC
phosphate buffer 202
phosphatidylinositol 88
phospholipid 89
  quantification 89
phosphoprotein 31, 36, 185
  stability 185
phosphorescense 87
photographic reducer 61
physical constants 211
PicoGreen 15
PITC 83
Pluronic F-68 227
PMSF 165, 169, 215, 221
pNP *see* p-nitrophenyl phosphate
polyethylene glycol *see* PEG
polyethyleneglycol 146, 151
polystyrene 156
polyvinylpyrrolidone 71, 76
Ponceau S 64, 99
positive predictive value 244
potassium dodecylsulfate 29
pre-flashing 80
pre-immun serum 72, 144, 153
precipitation aid 153
precipitin 151
propidium iodide 46
protein
  electroelution 66
  precipitation 67
Protein A 118, 145, 146, 150, 153
  binding affinities 147
Protein G 145, 150, 153
  binding affinities 147
protein kinase 185
Protein L 145, 146, 153
  binding affinities 147
PVDF 156
Pyronin Y 27, 37, 39, 49

quaternary structure 38

radioactivity 182
  half-life 182
  specific 183
radioisotope 182
radius, hydrodynamic 97
rcf 161
Reactive Red 121
receptor 170, 237, 238
  dihydropyridine 170, 173
recombinant protein 123
reference number 222
refractive index 228
regression 235, 236
  linear 235
  non-linear 236, 240
rehydration buffer 43
Reinheitszahl 135, 137
relative centrifugal field *see* rcf
R_f value 243
ribose 19
rotor 162
  fixed-angle 162
  swinging-bucket 162
  vertical 162, 177
rp-HPLC 108
RPMI 1640 205, 206

sample applicator 43
sarcolemma 167
sarcoplasmic reticulum 165
SCATCHARD plot 175, 238
SCHIFF bases 135, 137
SCHIFF's reagent 60, 62
scintillation cocktail 189
scintillator 81
SDS 6, 23, 226
sensitivity 244
serine phosphate 49

SI prefixes 210
SI units 210
significance 237
SMCC 130, 131, 133, 134
SOUTHERN blot 68, 78
spacer 110, 115, 116
specificity 244
SSC 204
stabilizator 91
stacking gel 27
staining, second 57, 63
standard curve 20
standard deviation 233
standard error 233
steady state 241
streptavidin 76, 121, 143, 158
streptomycin 179
Sulforhodamin 65
sulfosalicylic acid 124
SYBR Green 46

tannic acid 141
TBS 204
TCA 124
TCEP 26
TCEP · HCl 58
TE buffer 15, 204
thiocyanate 108, 121
thiophilic chromatography 147
threonine phosphate 49
thyreoglobulin 130
TLCK 221
TMB 73, 158, 159
  precipitating 157
total nitrogen determination 10
TPCK 221
TRAUT's reagent 132
trehalose 126
triazine dye 121
trichloroacetic acid *see* TCA
TRITC 139, 140
Triton X-100 6, 37, 227
Trypan blue 179
tungstate, sodium 124
Tween 20 227
Tween 80 227
Tyrode Solution 206
tyrosine phosphate 49

urea 34, 38, 47, 108, 121

volatile buffer 207

Western blot 41, 45, 68, 130
Western blotting 30, 36
wheat germ agglutinin 75, 117

X-phosphate 73

Zwittergent 227

Printing: Krips bv, Meppel
Binding: Stürtz, Würzburg